PIPE FITTER'S MATH GUIDE

E FITTER'S MATH GUIDE

PIPE FITTER'S MATH GUIDE

PIPE FITTER'S MATH GUIDE

PIPE FITTER'S MATH GUIDE

PIPE FITTER'S MATH GUIDE

PIPE FITTER'S MATH GUIDE

PIPE FITTER'S MATH GUIDE

PIPE FITTER'S MATH GUIDE

PIPE FITTER'S MATH GUIDE

PIPE FITTER'S MATH GUIDE

PIPE FITTER'S MATH GUIDE

PIPE FITTER'S MATH GUIDE

PIPE FITTER'S MATH GUIDE

PIPE FITTER'S MATH GUIDE

PIPE FITTER'S MATH GUIDE

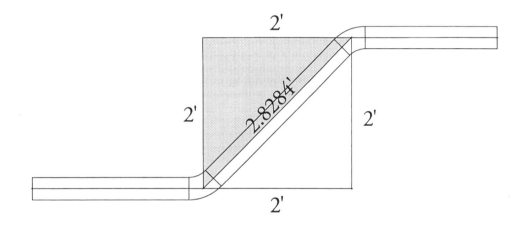

Johnny E. Hamilton
Pipe Fitter

Construction Trades Press
2265 Southeast Boulevard
Clinton, NC 28328

Pipe Fitter's Math Guide Copyright © 1989 Johnny E. Hamilton

For information:
Construction Trades Press
2265 Southeast Boulevard
Clinton, NC 28328

Printed in U.S.A.

Interior Design by Margaret Hamilton
Cover Design by Nikolas Wiltse

Website:
http://www.pipefitter.com/

Library of Congress Cataloging - in - Publication Data

Hamilton, Johnny E.,   1947-
    Pipe fitter's math guide   / Johnny E. Hamilton
    p.   cm.
    ISBN 0-9624197-0-2 (v.1)
    1. Pipe-fitting-- Mathematics   I. Title
    TH6715.H36   1989
    621.8'672'0151-- dc20   89-92081
                                             CIP

The ASME B16.9 standard changed ¾" 90° elbows Take Out to 1 ½" and the 45° elbows Take Out to ¾" in 2007.

26 25 24 23

Table of Contents

This book is dedicated to the people who teach with their lives.

Preface

As a pipe fitter, I was always looking for a book that would explain pipe math in a straightforward manner. I wasn't interested in the concepts of math, but in solutions to the practical problems that I faced on the job. I never found that book, but I did find the information I was seeking in college math classes. Over the years, other fitters have asked me how to do certain fits. While trying to explain, I realized that many were lacking the math skills needed; so I started writing the information down for them, and thus this book started to evolve.

Most pipe fitters are well trained in the practical aspects of the job, but lack the needed math skills. It is an insecure feeling to know that there are fits that you might be asked to do and yet you don't have the skills to approach them. There are tricks of the trade to help with some of them, but there are other fits that those short cuts won't cover. In those cases, you can only resort to the dreaded stovepiping. Stovepiping is a fancy term for guess work. There is an unwritten rule that you never say you can't do something. You eyeball the situation, and then fab a piece, put it up, cut it down, trim it, put it back up, and eyeball it again. It takes a long time, and the welder is eyeballing you while you are doing all of this, wondering if he can get another fitter. The only way to be secure in your work is to increase your skill level in math to the point that you have the tools necessary to do your job. A book to help me learn these skills is the one I was looking for, and now you have such a book.

This is not just a book for journeymen. It is for anyone who wants to gain more knowledge and confidence in the pipe trades. It is for the apprentice who wants to learn and move up as quickly as possible. It is for the supervisor who the journeymen and the apprentices go to for help in learning their trade. It is for the welder who wants to make the transition to fitter or foreman.

There will be those who say, "I don't need this math to get a job," but how much longer will that be true? More and more often management is asking, "Can you figure a rolling offset?" "Can you fab a reducer out of pipe?" Employers are getting serious about having well trained employees. They are beginning to realize how much it can cost them to have untrained people on the job.

There are people willing to put in the time it takes to know their trade. They know that with the proper training they will be faster and more efficient. Employers are beginning to see that they can afford to pay more for people who can work more efficiently.

This book is written to help you learn the math skills you need to be more efficient. If you are willing to work at it and invest in a scientific calculator, then this book will help you. It will teach you about right triangles and how to calculate their sides and angles. It will show you how to calculate the arc lengths of circles, and how to mark a flange by knowing just the number of bolt holes and the diameter of the bolt hole circle. It will help you in marking the arc lengths on odd angle elbows and to find the take out for those elbows.

There are 41 sets of practice questions with the answers in the back. You can work them on your own, get a group together, or perhaps talk to your employer about getting an instructor to teach a class. These are the first two modules, and there are more to come.

One last thing: don't look for short cuts. I've given you all the short cuts you need. Short cuts are fine when you know the entire process you are shortening, but not until then. Relax and take it one bit at a time. From the very beginning, you will see that this is usable knowledge that will help you do your job and help you gain confidence in your ability as a pipe fitter.

A note about scientific calculators: They are the right tool for the job and are relatively inexpensive these days. I use a Casio 991N that I picked up at one of the discount stores for $16, and I don't use all the keys on it. I suggest that you buy one and get familiar with how it works before you start this book. Some calculators require that you push a combination of keys to obtain what I say requires one key. This sort of difference is something you will find out by reading the calculator manual. Whichever calculator you buy, check to make sure it has the following keys.

| sin | cos | tan | sin⁻¹ | cos⁻¹ | tan⁻¹ | $\frac{1}{x}$ | π | x² | √x |

Introduction Addedum (2004)

One nice thing about writing a math book is that the math itself never changes, so the book doesn't have to be rewritten every two years to keep up with the times. Not so with calculators. Since the copyright date of this book, a new style of calculator has been introduced into the market which reacts differently than the initial examples in this book show.

If you buy a calculator today, it could be either type. The new format accepts information more like we speak. For example: We say sine of 45°, not 45° sine.

New Style: Press the function key (sine), enter 45, then push the = key.

Standard: Enter 45, then push the sine key.

For our purposes, we will refer to this new style of calculator as a **F**unction **F**irst calculator. To help the users of the FF calculator, we have incorporated new instructions and a new icon in the book. When the users of the FF calculators see the icon FF/Page 33 , they should go to the page indicated and work that example instead of the original. If you have a FF calculator that reacts other than the way I describe, don't be surprised. It could be just a difference in the brand of calculator. Use the book that came with the calculator to figure out the right method for your calculator.

Which style calculator do you have? You can quickly tell the difference by keying in 45, then pushing the sin button. If you have the FF style, it will display a 0 (zero) or an error. The standard style will display 0.707106781, which is the sine for 45°. I suggest for ease of use that you buy the standard type of calculator.

You are learning a new skill, and with each new skill there will be new terms. Some of these terms you may already know, and some you may not. This is the language of math, and it is important to know, just as our language in pipe is. Using these terms will increase your communications skills with the engineers and designers since you will be speaking the same language. Throughout the book, when a term is used that you don't understand, look it up. The glossary in the back of the book will give you a short definition, if you need more, the index can refer you to a page number that gives a more detailed explanation.

Acknowledgements

There are a lot of people to thank for helping me.

My wife, Margaret, who worked and suffered through this with me. She and Marchelle made do with less so this book could be written. They would appreciate it, if you would destroy any unauthorized copies you see. My parents, W.E. and Frances Hamilton,who raised me to think I could do the things I really wanted to. My brothers, Ralph and James, who also work as pipe fitters, for urging me on and for being my friends. My sisters, Eleanor and Billie, who taught me manners I sometimes use, and who also gave me moral support. Thank you, Ruby Sutton for being an understanding mother-in-law and thank you, Esther Best for a lifetime of support.

I also want to thank Leer Larkins, Nosotros Workshop, Zacatecas, the Medinas, and the Sandells in Riverside, California, for all the help they have given me through the years. Thank you, Ted and Linda Jones, and Thomas and Liz Boykin. Thank you, Nick, for telling me to use what I have instead of always trying to learn more. Thanks, Gibbs, for wanting to know more.

Kenneth Jones was kind enough to read through the book to make sure I didn't take too many liberties with the rules of math. He is a well respected instructor at Sampson Community College in Clinton, N.C. and at Fayetteville State University in Fayetteville, N.C.

You would not be able to read this without **Chris and Rebecca Forhan**. Chris is an English instructor at Trident Technical College in Charleston, S.C. He had the task of making this book readable. Rebecca is a painter and she helped make it pleasing to the eye.

Any mistakes you find are mine. I am still making changes and will do so until we go to press.

Thank you for buying this book, and I'll see you down the road.

Module One

Pipe Math

Answers to the practices are located on pages 131-138. If you find that your answers do not match, study the examples again. If you still have a problem understanding a section after a few tries, ask someone for help. It is important to be able to correctly work the practices in one section before attempting the next.

I

Fractions and Decimals

There are many units of measurement that we use every day to indicate distance, such as inch, foot, yard, mile, and meter; however, in the pipe fitting trade we generally only use the inch and the foot. The accuracy of our fits requires that we be able to gauge the distances that fall in between the whole inches. A machinist may require that an inch be divided into a thousand or ten thousand equal parts for the accuracy that he needs, but in pipe fitting, dividing the inch into 16 equal parts is usually satisfactory. The distance that falls between the whole units, or inches in this case, is expressed by what is called a *fraction*.

A fraction indicates that you are dealing with a part of a whole. Just as one quarter ($.25) indicates a part of a whole unit (the dollar), one quarter of an inch indicates a part of a whole unit (the inch). In both of the above fractions you are dealing with one part of a unit that has been divided into 4 parts. It takes 4 quarters to make one dollar, and it takes four $\frac{1}{4}$" to make one inch.

Example: $2\frac{1}{2}$" ($2\frac{1}{2}$ inches) falls between 2" and 3"

2 and 3 are the whole numbers

$\frac{1}{2}$ is the fraction

Inch is the unit

The bottom number of a fraction is called the denominator and indicates the number of parts the whole has been divided into. The top number, called the numerator, indicates the portion of the whole that you are dealing with.

Numerator ➡ 1
Denominator ➡ 4

Different fractions can be equal to each other; one quarter equals $\frac{1}{4}$ of a dollar, but so does 5 nickels or 25 pennies.

$$\frac{1}{4} = \frac{5}{20} = \frac{25}{100}$$

Just as you have less bulk to carry around when you have a quarter in your pocket instead of 25 pennies, fractions are easier to deal with when you reduce them to their lowest term.

To reduce a fraction to its lowest term, you must be able to divide the numerator and the denominator by the same number, with the answers for both being whole numbers. For example, in the fraction $\frac{5}{20}$ both the numerator and the denominator can be divided by 5.

$$\frac{5}{20} = \frac{5 \div 5}{20 \div 5} = \frac{1}{4}$$

If the top and bottom numbers can be divided by several numbers, use the largest possible number as the divisor (divider).

Example: $\frac{25}{100}$ can be divided by 5 or 25. In both cases, the answers for both the numerator and denominator will be whole numbers, but you will save yourself a step by using the larger number.

$$\frac{25}{100} = \frac{25 \div 5}{100 \div 5} = \frac{5}{20} = \frac{5 \div 5}{20 \div 5} = \frac{1}{4}$$

or

$$\frac{25}{100} = \frac{25 \div 25}{100 \div 25} = \frac{1}{4}$$

> When the numerator and the denominator of a fraction are both even numbers, the fraction can always be reduced to a lower term by dividing by 2.

Practice 1: Reduce these fractions to their lowest term.

(1) $\frac{6}{16}$ (5) $\frac{12}{16}$

(2) $\frac{2}{8}$ (6) $\frac{2}{4}$

(3) $\frac{4}{16}$ (7) $\frac{14}{16}$

(4) $\frac{24}{32}$ (8) $\frac{6}{8}$

Converting Fractions to Decimals

Fractions are used when measuring distances in the field; the fractions are then converted to decimals in order to perform the calculations. After the calculations, fractions are again needed for measurements, so the decimals are converted back to fractions. The conversions are necessary because of the differences between the measuring tools and the calculating tools.

We use conversions of fractions and decimals every day. Our money system is based on the decimal system.

$\frac{1}{2}$ of a dollar is equal to $.50 (a half dollar)

$\frac{1}{4}$ of a dollar is equal to $.25 (quarter)

$\frac{1}{8}$ of a dollar is equal to $.125 ($12\frac{1}{2}$ cents)

$\frac{1}{10}$ of a dollar is equal to $.10 (dime)

Just as fractions of a dollar can be converted into decimals, fractions of an inch can be converted into decimals:

$\frac{1}{2}$ of an inch is equal to .50"

$\frac{1}{4}$ of an inch is equal to .25"

$\frac{1}{8}$ of an inch is equal to .125"

$\frac{1}{10}$ of an inch is equal to .10"

Decimal place values

$.1" = \frac{1}{10}$ "

$.01" = \frac{1}{100}$ "

$.001" = \frac{1}{1000}$ "

$.0001" = \frac{1}{10,000}$ "

$.00001" = \frac{1}{100,000}$ "

Converting Fractions of Inches to Decimal

The bar between the top number (numerator) and the bottom number (denominator) of a fraction is a division bar.

$$\frac{\textbf{numerator}}{\textbf{denominator}}$$

It is easy to convert a fraction to a decimal using a calculator. All you have to do is divide the numerator by the denominator. The answer will be the decimal which is equal to that fraction.

Example: $\frac{9}{16}$" = .5625"

Enter 9 $\div$ 16 $=$ display | 0.5625 |

You can use a conversion chart, but you will save yourself a lot of time in the future by learning to do the conversions on the calculator.

Practice 2: Convert these fractions of inches to decimals by division. The answers are on page 131.

(1) $\frac{3}{8}$"

(2) $\frac{11}{16}$"

(3) $\frac{7}{16}$"

(4) $\frac{1}{2}$"

(5) $\frac{3}{4}$"

(6) $\frac{19}{32}$"

(7) $\frac{7}{8}$"

(8) $\frac{3}{16}$"

(9) $\frac{1}{8}$"

(10) $\frac{1}{4}$"

(11) $\frac{15}{16}$"

(12) $\frac{1}{16}$"

Converting Mixed Numbers to Decimals

When converting a mixed number (a whole number and a fraction) to a decimal, you must first subtract the whole number from the mixed number. Next, convert the fraction to a decimal, and then add the whole number back to the decimal.

Example: $7\frac{7}{16}$" is a mixed number

$7\frac{7}{16}$" - 7" = $\frac{7}{16}$"   (subtract the whole number from the mixed number)

$7 \div 16 = .4375$"   (convert fraction to decimal)

$.4375$" + 7" = 7.4375"   (add the whole number to the decimal)

Practice 3: Convert these mixed numbers to a whole number with decimals.

(1) $17\frac{3}{4}$"   (5) $7\frac{1}{8}$"

(2) $43\frac{15}{16}$"   (6) $87\frac{7}{8}$"

(3) $23\frac{11}{16}$"   (7) $2\frac{1}{32}$"

(4) $3\frac{3}{16}$"   (8) $4\frac{5}{8}$"

Converting a Fraction of a Foot to a Decimal

Many times the measurements that are taken in the field are in feet rather than inches. In these cases, you again need a fraction so that you can divide the numerator by the denominator in order to convert to a decimal.

Remember, the denominator is how many parts a unit has been divided into. In this case there are 12 inches to a foot, so 12 is the denominator. The numerator indicates the number of parts you are dealing with; therefore, the number of inches in the measurement is the numerator. The example below converts whole inches to a fraction of a foot.

Remember that the bar between the numerator and denominator is a division bar.

Example:

Measurement		Fraction		Decimal
9'3"	=	$9\frac{3}{12}$'	=	9.25'
2'6"	=	$2\frac{6}{12}$'	=	2.5'
10' 11"	=	$10\frac{11}{12}$'	=	10.9167'*

Practice 4: Convert these measurements to decimals (first, create a fraction). Round off to the fourth decimal place.

(1)   5' 7"
(2)   4' 1"
(3)   92' 5"
(4)   19' 9"
(5)   46' 10"
(6)   137' 8"

Converting a mixed number of feet, inches, and a *fraction of an inch* to feet and decimal of a foot requires an extra step. You have to convert the fraction of an inch to a decimal of an inch. Look at the example below to see what a fraction of a foot that includes a fraction of an inch looks like.

$$5' 4\frac{3}{8}" = 5\frac{4\frac{3}{8}}{12}' = 5\frac{4.375}{12}'$$

The conversion is completed by dividing the 12 (the number of inches in a foot) into 4.375.

Here is a series of steps for the conversion.

First   convert the fraction of an inch to a decimal of an inch.

Second   add the whole inches to the decimal of an inch.

Third   divide the new number by 12 (the number of inches in a foot).

Finally   add in the whole feet.

---

* In the case of this decimal, I have rounded off the decimal because 4 decimal places is generally accurate enough for the pipe fitter's purpose. Rounding a number off is done by looking at the number to the right of the number to be rounded off. If the number to the right is 5 or higher, then the round off number is made one number larger. If the number to the right is less than 5, then the round off number stays the same. In both cases, the round off number is the last number in the series. In the example shown above, 10.91666666 rounds off to 4 decimal places to become 10.9167. The 6 is the fourth decimal place and the number to the right of it is a 6, so the fourth decimal place is rounded up a number to 7, and all of the numbers to the right of the fourth decimal place are dropped.

Example:

73' 7 $\frac{13}{16}$" (expressed as a fraction of a foot: 73$\frac{7\frac{13}{16}}{12}$')

Divide the fraction of the inch: 13 ÷ 16 = .8125"

Add in the whole inches: 7" + .8125" = 7.8125"

Divide by the number of inches in a foot: 7.8125" ÷ 12" = .6510'

Add the whole feet: 73 + .6510 = 73.6510'

Here is the same example done on the calculator.

Enter		Display	
13 ÷ 16 =	Display	0.8125
+ 7 =	Display	7.8125
÷ 12 =	Display	0.651041666
+ 73 =	Display	73.65104167

73.6510' is the answer after rounding off.

Practice 5: Convert these measurements to decimals of a foot (first, create a fraction). Round the answers off to 4 decimal places.

(1) 6' 6"

(2) 1' 10"

(3) 3' 5 $\frac{3}{4}$"

(4) 16' 8 $\frac{1}{16}$"

(5) 23' 4 $\frac{5}{8}$"

(6) 123' $\frac{3}{8}$"

(7) 42' 9 $\frac{9}{16}$"

(8) 1' 11 $\frac{13}{16}$"

(9) 96' 3 $\frac{1}{4}$"

(10) 37' 5 $\frac{15}{16}$"

Converting Decimals to Fractions

In the field, calculations are made to obtain a needed measurement. Decimals are used in calculations, but measurements are made in fractions, so you will convert your final answers from decimals to fractions.

Converting a Decimal of an Inch to a Fraction

To convert a decimal of an inch (.375) to a fraction of an inch $(\frac{3}{8})$, simply reverse the method you used to convert a fraction to a decimal. Multiply the decimal of an inch by the denominator that you will be using. Since the accuracy generally needed in our trade is based on $\frac{1}{16th}$ of an inch, 16 is the most often used denominator. If more accuracy is needed, 32 can be used. The nearest whole number of the answer will be the numerator.

Example: .375   (the decimal of an inch)
              x 16   (the denominator)
             6.000   (the numerator)

You now have the fraction $\frac{6}{16}$", which is reduced to $\frac{3}{8}$".

When working problems, seldom do you end up with a decimal that will convert to an even sixteenth. You have to round off the numerator to the closest whole number. A decimal such as .6327 multiplied times 16 gives an answer of 10.1232. The answer is the numerator and the multiplier is the denominator.

For example: .6327
                 x16   Numerator   10.1232
                 ─────   ─────────   ───────
                 10.1232   Denominator   16

Round off the numerator to the nearest whole number and the fraction reads $\frac{10}{16}$, and when reduced, $\frac{5}{8}$.

Remember: When rounding off to the nearest whole number, if the decimal is .5 or more you go up to the next whole number. If the decimal is below .5 you use the existing whole number. For example, 3.49 will round off to 3, and 3.5 will round off to 4.

Practice 6: Convert these decimals to the closest sixteenth. Reduce to the lowest fraction possible.

(1) .4375"   (5) .9426"

(2) .6250"   (6) .1743"

(3) .2912"   (7) .8267"

(4) .5555"   (8) .3164"

Notes

Converting a Number with Decimals to a Mixed Number

To convert inches with decimals (9.5) to inches and fractions ($9\frac{1}{2}$), you must first subtract the whole inches (9) so that you end up with just the decimal (.5) in the calculator. Since you are converting so you can make a measurement, it sometimes helps to write the whole number down when you subtract it out. Then write the fraction next to the whole inches.

Example: 9.1875"

9.1875" - 9" = .1875"   Write down ✍ 9" or $9\frac{}{16}$"

.1875 x 16 = 3   3 is the numerator 👉 $9\frac{3}{16}$"

Practice 7: Convert from a number with decimals to a mixed number. Reduce to the lowest term when necessary.

(1) 12.375"   (5) 74.1725"
(2) 4.6"   (6) 19.5737"
(3) 99.913"   (7) 37.4189"
(4) 56.78"   (8) 23.0892"

Converting a Decimal of a Foot to a Fraction

To convert a number with feet and a decimal of a foot (9.5') to a number with feet, inches, and a fraction of an inch, simply reverse the process used to find the decimal of a foot. Subtract the number of whole feet, then multiply the decimal of a foot by 12 (the number of inches in a foot). The whole number in the answer is the inches. It helps to write the numbers down while you are doing the conversion.

Write down ✍

6.75'- **6'** = .75   **6'**
.75 x 12" = **9"**   6'**9"**

Most of the time when you multiply by 12, the answer is a whole number and a decimal. That decimal is the decimal of an inch, and the whole number is the whole inches. To change the decimal of an inch to a fraction of an inch, subtract the whole number and multiply the decimal times 16.

Write down ✍

Example: 73.65104' - **73'** = .65104'   **73'**

.65104 x 12 = 7.81248" - **7"** = .81248   73'7"

.81248 x **16** =12.9996 (round off to **13**)   73' 7 $\frac{13}{16}$"

Practice 8: Convert these whole feet with decimals to feet, inches, and fractions of inches. Reduce the fractions when necessary.

(1)   24.75'   (7)   33.894'

(2)   1.9'   (8)   75.128'

(3)   3.333'   (9)   84.439'

(4)   170.863'   (10)   4.567'

(5)   53.125'   (11)   8.298'

(6)   42.7429'   (12)   12.997'

The key to using this book well is being aware of what you know and what you don't know. Just how much have you learned from your work so far? Many people in school learn just enough to repeat it back to the teacher or to pass a test. In this case, that is not enough. Some of you will be seeing new material in the coming sections and you don't need to be struggling with converting fractions and decimals while trying to learn something new. You need to know the previous sections backwards and forwards. Below is a self test to help you determine if any of the previous sections need to be worked again or if you are ready to move on. The answers can be found on page 138.

Treat the numbers below as either measurements that need to be converted to decimals for use in the calculator or answers from the calculator that need to be converted to allow you to measure in the field.

Remember to look at the sign for feet ' or the sign for inches " to determine which way you approach each problem.

(1)	11.6485"	(11)	93.7564'
(2)	5.9624'	(12)	$23\frac{7}{8}$"
(3)	42' 3"	(13)	72.5649"
(4)	$14\frac{13}{16}$"	(14)	$47' 6\frac{1}{16}$"
(5)	16.7437'	(15)	89.6789'
(6)	$28' 5\frac{3}{16}$"	(16)	56.4569"
(7)	10.123"	(17)	1.0032'
(8)	$33\frac{3}{4}$"	(18)	$4\frac{3}{8}$"
(9)	3.345'	(19)	24.9643"
(10)	$148' 11\frac{5}{8}$"	(20)	$47' 5\frac{7}{16}$"

If you worked through these problems with confidence, move on. If you felt a little awkward, work the previous sections again. In fact, throughout the book make it a habit to proceed only when you feel confident.

Memory Aid

Doubling a Fraction

To double a fraction, you normally multiply it times 2, then reduce to the lowest denominator.

Example:

$$2 \times \frac{3}{8} = \frac{2}{1} \times \frac{3}{8} = \frac{2 \times 3}{1 \times 8} = \frac{6}{8} = \frac{3}{4}$$

A quicker way is to divide the denominator by 2.

$$\frac{3}{8 \div 2} = \frac{3}{4}$$

Of course, this works only with fractions that have an even number for the denominator, but a pipe fitter's measuring fractions are all based on even numbers. (Examples; $\frac{1}{2}, \frac{1}{4}, \frac{1}{8}$, and $\frac{1}{16}$.)

Halving a Fraction

To divide a fraction in half, we were told in school to invert and multiply. The shortcut is just to multiply the denominator by 2.

To find half of these fractions, multiply the denominator by 2.

$$\frac{3}{8} \div 2 = \frac{3}{8 \times 2} = \frac{3}{16} \qquad \frac{7}{8} \div 2 = \frac{7}{8 \times 2} = \frac{7}{16}$$

$$\frac{3}{4} \div 2 = \frac{3}{4 \times 2} = \frac{3}{8} \qquad \frac{5}{8} \div 2 = \frac{5}{8 \times 2} = \frac{5}{16}$$

Halving Mixed Numbers

A few years ago I learned a new trick from an old Bulldog, Bulldog Holland, on how to divide a mixed number in half. He separated mixed numbers into two categories. The first category was mixed numbers with even whole numbers, and the second category was mixed numbers with odd whole numbers.

For example, $32\frac{7}{8}$ belongs in the first category (mixed numbers with even whole numbers).

In this case, divide the whole number by 2: $\frac{32}{2} = 16$

Use the short cut for dividing a fraction: $\frac{7}{8 \times 2} = \frac{7}{16}$

Add them back together: $16\frac{7}{16}$

Half of $32\frac{7}{8}$ is $16\frac{7}{16}$

$11\frac{5}{8}$ belongs in the second category (mixed numbers with odd whole numbers).

In this case, divide the whole number by 2: $\frac{11}{2} = 5.5$

Drop the remainder to make it a whole number of 5.

Add the numerator to the denominator: $5 + 8 = 13$ (13 is the new numerator).

Then double the denominator: $8 \times 2 = 16$ (16 is the new denominator).

When you put the two together, the fraction is $\frac{13}{16}$.

Your final answer is $5\frac{13}{16}$.

This drawing shows how it works.

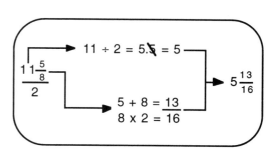

II

Angles

Angle is a very important term and has many definitions. The three definitions below will cover most of what we need.

1. The shape made by two straight lines meeting in a point.

2. The space between those lines.

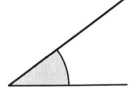

3. The amount of space measured in degrees.

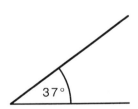

Vertex is the point where the two straight lines come together to form the angle.

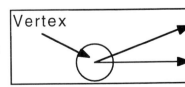

Angle Names

Names are given to angles with certain characteristics. These names will be used throughout the book.

Zero Degree

A **zero degree angle** has no space between the two lines.

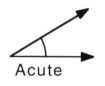
Acute

An **acute angle** is any angle between 0° and 90°.

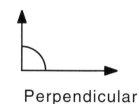
Perpendicular

A **perpendicular** has a 90° angle between the two lines. It is also called a **right angle**.

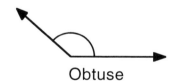
Obtuse

An **obtuse angle** is an angle between 90° and 180°.

Straight

A **straight angle** is one of 180°.

A **central angle** is an angle in which the vertex is located in the center of the circle.

III

Degrees

Degrees are the units of measure for angles. One degree is equal to $\frac{1}{360}$ of the complete rotation of a circle. The center of the circle is the angle's vertex.

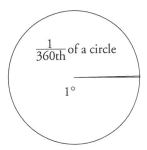

You shouldn't be surprised to find out that there are 360° in a circle.

Each degree can be divided into 60 minutes. 1° = 60'.
Each minute can be divided into 60 seconds. 1' = 60".
The symbol for: degrees is °, minutes is ', seconds is ".

Degrees, minutes, and seconds can also be expressed as degrees and a decimal of a degree. An example is 37° 42' 17" expressed as 37.7047°.

Most scientific calculators can express degrees in both ways. The key for degrees on my calculator looks like $\boxed{°\ '\ "}$, but the key on another brand may look like $\boxed{\text{DMS}}$. You will need to refer to your calculator manual to determine the correct key for degrees. Most calculators display answers in the form of degrees and a decimal of a degree. Unless you are just curious, there is little reason to convert this figure to DMS (degrees, minutes, seconds).

Degrees and Circles

A circle has 360°, a half circle has 180°, a quarter circle has 90°, and an eighth of an circle has 45°.

It is important to understand how a circle is broken down when cutting elbows. We will cover this in more detail in another section.

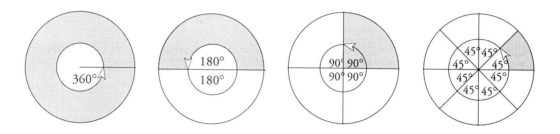

Degrees and Triangles

The sum of the three angles which make up any triangle will *always* equal 180°.

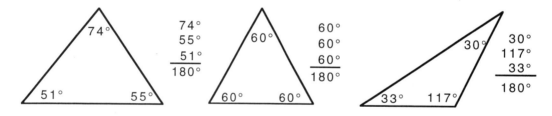

A right triangle will *always* have one angle equal to 90° and two angles whose sum is 90°, making a total of 180°.

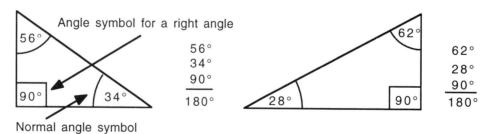

IV

Squares

A number is squared by multiplying it times itself. On the calculator, x^2 will **square** whatever is in the display.

Example

Enter   3   Display $\boxed{ 3}$

$\boxed{x^2}$ Display $\boxed{ 9}$

Here are some examples of squares:

$$
\begin{aligned}
1^2 &= 1 \times 1 &= 1 \\
2^2 &= 2 \times 2 &= 4 \\
3^2 &= 3 \times 3 &= 9 \\
4^2 &= 4 \times 4 &= 16 \\
5^2 &= 5 \times 5 &= 25 \\
6^2 &= 6 \times 6 &= 36 \\
7^2 &= 7 \times 7 &= 49 \\
8^2 &= 8 \times 8 &= 64 \\
9^2 &= 9 \times 9 &= 81 \\
10^2 &= 10 \times 10 &= 100 \\
a^2 &= a \times a &= a^2 \\
b^2 &= b \times b &= b^2 \\
c^2 &= c \times c &= c^2
\end{aligned}
$$

The last three examples use a letter instead of a number. Those letters are called *variables*. Variables represent any number or an unknown number. In this case they show us the rule that any number squared, is the number times itself.

Most of the numbers you will be squaring will be mixed numbers such as inches and fractions and feet, inches, and fractions. You will have to convert the fractions to decimals then square.

Example: What is the square of $4' 9\frac{7}{16}"$?

$$(4' 9\tfrac{7}{16}")^2 = (4\frac{9\frac{7}{16}}{12})^2 = (4\frac{9.4375}{12}')^2 = (4.786456333)^2 = 22.9102$$

Enter	$7 \div 16$	Display		.4375
	$+ 9 =$	Display		9.4375
	$\div 12 =$	Display		.786458333
	$+ 4 =$	Display		4.786458333
	x^2	Display		22.91018338

Practice 9: Use the x^2 key to find these squares. Round off to 4 decimal places.

(1) 10^2 (5) 5.5625^2

(2) 23^2 (6) $(9\tfrac{3}{8})^2$

(3) 230^2 (7) 11^2

(4) 17.5^2 (8) 25^2

V

Square Roots

Square root poses the question: what number, times itself, will equal this number? Another way of asking the same question is: what number has been squared to get this number? Our best known square root, $\sqrt{2}$, is 1.4142*, and is used to calculate a 45° offset. The $\boxed{\sqrt{x}}$ key will find the square root of whatever is in the display on your calculator.

Example: Find the square root of 16. **FF/Page34**

16 $\boxed{\sqrt{x}}$

Display | 4 |

You can check yourself by squaring the answer.

4 $\boxed{x^2}$

Display | 16 |

Here are some examples of square roots.

$$\sqrt{1} = 1$$
$$\sqrt{4} = 2$$
$$\sqrt{9} = 3$$
$$\sqrt{16} = 4$$
$$\sqrt{25} = 5$$
$$\sqrt{36} = 6$$
$$\sqrt{49} = 7$$
$$\sqrt{a^2} = a$$
$$\sqrt{b^2} = b$$
$$\sqrt{c^2} = c$$

The last three examples use variables instead of numbers. These examples show another rule of math, the square root of a number squared is the number. For example: $\sqrt{4}$ or $\sqrt{2^2} = 2$.

Every number has a square and a square root.

_____

* This square root has been rounded off to 4 decimal places.

Practice 10: Find the square root of the following numbers using the $\boxed{\sqrt{x}}$ key. Round off to 4 decimal places.

(1)	2	(5)	256
(2)	69	(6)	1037
(3)	14.5	(7)	94.5625
(4)	$17\frac{3}{4}$	(8)	$34\frac{7}{8}$

$\boxed{FF}$ **Note:** The sentence on Page 33 that reads, "The $\boxed{\sqrt{x}}$ key will find the square root of whatever is in the display on your calculator." is incorrect for FF calculators. For Function First calculators you must push the $\boxed{\sqrt{x}}$ key, enter the number, then push =.

For FF type calculators, the example on the page 33 should be entered like this:

Example: Find the square root of 16.

$\boxed{\sqrt{x}}$ Enter 16   Display | $\sqrt{16}$ |

$\boxed{=}$ Display | 4 |

If you want to check yourself, push $\boxed{x^2}$ while the 4 is still in the display and you will see 16.

Note: Of the calculator keys shown on the Preface page only the $\boxed{x^2}$ and the $\boxed{\frac{1}{x}}$ will react to what is in the display for the **FF** style calculators.

Notes

VI

Right Triangles

$$a^2 + b^2 = c^2$$

This formula* will allow you to verify that a triangle is a right triangle, and to calculate the length of the third side of a right triangle when the other two sides are known. The letters a, b, and c are called **variables**, and they represent the lengths of the three sides of a **right triangle**.

The variable c is *always* used to represent the **hypotenuse**. The hypotenuse is *always* the longest side and is *always* directly across from the 90° angle. The other two sides are often called the **legs**.

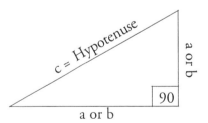

At this point, it really doesn't matter which one of the legs you call <u>a</u> or <u>b</u>, as long as you make one <u>a</u> and the other <u>b</u>.

---

* This is the formula from the Pythagorean Theorem. The theorem says that the square of the hypotenuse is equal to the sum of the squares of the other two sides in a right triangle.

The 3-4-5 right triangle is often used because it is simple to work with. Let's use this triangle to show how the formula works to verify that this is a right triangle.

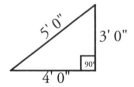

$a = 3$ $b = 4$ $c = 5$

$$a^2 + b^2 = c^2$$
$$3^2 + 4^2 = 5^2$$
$$9 + 16 = 25$$
$$25 = 25$$

Since the numbers on both sides of the = mark are the same, the 3-4-5 triangle is a right triangle. Please note: This is a rule! Any triangle that can be worked this way is a right triangle.

Finding the Length of the Hypotenuse

In this example, we know the length of the legs a and b, but not the length of the hypotenuse. The original formula is used, but in a different format.

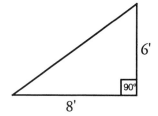

The formula for determining the length of the hypotenuse when the length of the two legs are known is

$$\sqrt{a^2 + b^2} = c.$$

Replace the variables with the lengths of the legs, do the calculations, and the result is the length of the hypotenuse.

$$\sqrt{a^2 + b^2} = c$$
$$\sqrt{6^2 + 8^2} = c$$
$$\sqrt{36 + 64} = c$$
$$\sqrt{100} = c$$
$$10 = c$$

Warning! Do not add $6^2 + 8^2$ to get 14^2. It doesn't work!

Here is how to work this formula on the calculator. 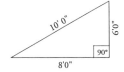 *Remember* that the keys x^2 and $\sqrt{x}$ react to whatever is on the display when they are pushed.

 **FF/Page 38**

Enter 6 x^2 +   display | 36

Enter 8 x^2 display | 64

=   display | 100

$\sqrt{x}$ display | 10

10 is the length of the hypotenuse.

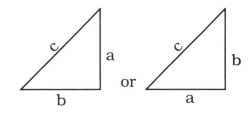

Practice 11: Find the length of the hypotenuse using the two sides given. Round off to 4 decimal places.

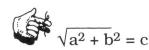 $\sqrt{a^2 + b^2} = c$

If you have problems squaring the mixed numbers refresh your memory on page 32.

(1) a = 9   b = 12   (7) a = 8' 3$\frac{3}{4}$"   b = 11' 11$\frac{11}{16}$"

(2) a = 15   b = 20   (8) a = 16.73"   b = 27.32"

(3) a = 21"   b = 33"   (9) a = 103' 3"   b = 76' 7"

(4) a = 2.5"   b = 7.75"   (10) a = 18   b = 24

(5) a = 9$\frac{3}{8}$"   b = 6$\frac{7}{16}$"   (11) a = 17$\frac{15}{16}$"   b = 23$\frac{15}{16}$"

(6) a = 7' 9"   b = 3' 4"   (12) a = 24   b = 45

FF **Note:** For FF calculators users only, the example on page 49 should look like this:

Push $\sqrt{x}$ Display [$\sqrt{}$ ( ]

Enter 6 x^2 + 8 x^2 *   Display [$\sqrt{}$ (6² + 8²)]

Push =   Display [10]

The *Remember* statement on Page 37, "**The keys** x^2 **and** $\sqrt{x}$ **react to whatever is on the display when they are pushed.**", is only half correct for FF calculators. The x^2 will respond to whatever is in the display.

To determine the square root of number, push the $\sqrt{x}$ key, enter the number, then push the = key. This means to find the square root for an answer to a calculation, you must think ahead and push the $\sqrt{x}$ key first, enter your calculation, then push the = key. It will take a little practice, but you'll get use to it.

---

* You may or may not need to push the right parenthesis for this calculation. It depends on the calculator.

Finding the Length of a Leg

If you know the length of the hypotenuse and the length of one of the legs and want to know the length of the other leg, the right triangle formula can again be rearranged to find the answer. The variable that represents the unknown side will need to be on a side of the equation by itself. Since you are naming the sides, it can be either $\underline{a}$ or $\underline{b}$.

The formula needed to find an unknown leg when the other sides are known is:

$$a = \sqrt{c^2 - b^2}$$

$$b = \sqrt{c^2 - a^2}$$

Notice that under the square root symbol, you subtract the square of the known leg from the square of the hypotenuse to find the unknown leg.

Example:

If the hypotenuse is 150 and one of the legs is 90, what is the length of the other leg? Let's say 90 is side $\underline{b}$.

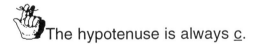The hypotenuse is always $\underline{c}$.

When we put these numbers into the formula, they look like this.

$$a = \sqrt{c^2 - b^2}$$
$$a = \sqrt{150^2 - 90^2}$$
$$a = \sqrt{22500 - 8100}$$
$$a = \sqrt{14400}$$
$$a = 120$$

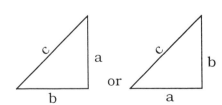

Here is how it is done with the calculator.

 FF/Page 40

Enter 150 x^2 $-$ display | 22500

Enter 90 x^2 display | 8100

       $=$ display | 14400

       $\sqrt{x}$ display | 120

120 is the third side, <u>a</u>.

Practice 12: Find the length of the third side. Round off to 4 decimal places.

  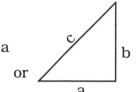

$b = \sqrt{c^2 - a^2}$

(1)	a = 12	c = 15	(7)	a = 7.375	c = 12.5625
(2)	a = $9\frac{1}{2}$	c = $17\frac{7}{16}$	(8)	a = 23.9375	c = 49.25
(3)	a = 7	c = 19	(9)	a = 4	c = 16
(4)	a = 5'2"	c = 9'5"	(10)	a = $10\frac{7}{8}$	c = 22
(5)	a = 78'	c = 100'	(11)	a = 5'	c = 10'
(6)	a = 14"	c = 24"	(12)	a = $1'\,0\frac{1}{4}"$	c = 2'

FF **Note:** For FF style calculators, you must press the $\sqrt{x}$ key, then enter your calculation.

Push $\sqrt{x}$ display | $\sqrt{\,}$ (

Enter 150 x^2 $-$ 90 x^2 *   display | $\sqrt{(150^2 - 90^2)}$

Push $=$ display | 120

---

* You may or may not need to push the right parenthesis for this calculation. It depends on the calculator.

Names of the Sides of a Right Triangle

Knowing the names of the sides of a right triangle is very important. Besides being a part of the vocabulary of math, these names can be used to communicate which side of a triangle you are referring to without using a diagram. You are about to study the ratios of these sides, and if you misname the sides your calculations will be off.

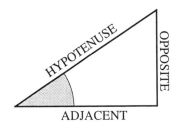

The gray shading indicates the reference angle.

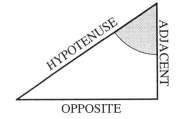

The **hypotenuse** is the side directly across from the 90° angle and is always the longest side of a right triangle.

The other two sides are assigned names according to the **reference angle***.

The side next to the reference angle is always the **adjacent side.**

The side directly across from the reference angle is always the **opposite side**.

It is important to realize that the names of the opposite and adjacent sides change when you move from one reference angle to the other.

---

* The reference angle is never the 90° angle.

Ratio of Sides

What is a ratio? A **ratio** is the comparison by division. You divide one factor into another for comparison. In sports, ratios are constantly flashed on the TV screen. In baseball, they indicate a player's number of times at bat compared to his number of hits. This ratio is called a batting average.

$$\frac{\text{Hits}}{\text{At Bats}}$$

In basketball, a ratio of shots attempted to shots made is called shooting average.

$$\frac{\text{Shots made}}{\text{Shots taken}}$$

A ratio of sides of a right triangle is determined by comparing the length of one of the sides of the right triangle to the length of another side.

$$\frac{\text{opposite side}}{\text{adjacent side}} \quad \text{or} \quad \frac{\text{hypotenuse}}{\text{opposite side}}$$

These ratios can be expressed as either fractions or decimals.

The ratios of the sides are directly related to the number of degrees in the reference angle. A relationship between the ratios of the sides and the number of degrees in a right triangle has been known for years. It was found that a right triangle that had the same angles as another right triangle also had the same ratio of sides. A system was worked out that enables us to compare the ratios to a chart (these days we use a calculator) and find the degrees of the angles. You can also use the same chart to find the ratio of sides for the different angles. What this means is that if you take the time to learn the names of the functions and the ratios that are assigned to them, you will be able to find the degrees of the angles of any right triangle just by knowing the length of two sides.

To introduce ratios, the 3-4-5 right triangle will be used again.

With each angle, there are only six possible ways that
the sides can be divided into each other, so there are only
six possible ratios for each angle. The six ratios for the
reference angle shown (indicated by the shading) are as
listed:

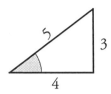

$$\frac{\text{Opposite}}{\text{Hypotenuse}} = \frac{3}{5} \qquad \frac{\text{Hypotenuse}}{\text{Opposite}} = \frac{5}{3}$$

$$\frac{\text{Adjacent}}{\text{Hypotenuse}} = \frac{4}{5} \qquad \frac{\text{Hypotenuse}}{\text{Adjacent}} = \frac{5}{4}$$

$$\frac{\text{Opposite}}{\text{Adjacent}} = \frac{3}{4} \qquad \frac{\text{Adjacent}}{\text{Opposite}} = \frac{4}{3}$$

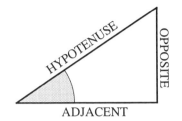

Each ratio has been assigned a
name.

These names are called **functions**.

Every angle has six functions.

The Functions

$$\text{Sine } \theta = \frac{\text{Opposite}}{\text{Hypotenuse}} \qquad \text{Cosecant } \theta = \frac{\text{Hypotenuse}}{\text{Opposite}}$$

$$\text{Cosine } \theta = \frac{\text{Adjacent}}{\text{Hypotenuse}} \qquad \text{Secant } \theta = \frac{\text{Hypotenuse}}{\text{Adjacent}}$$

$$\text{Tangent } \theta = \frac{\text{Opposite}}{\text{Adjacent}} \qquad \text{Cotangent } \theta = \frac{\text{Adjacent}}{\text{Opposite}}$$

This chart shows the functions and the ratios for this reference angle.
*The key to understanding right triangles is understanding how the ratios of
sides, the functions, and the angles are related.* θ represents the unknown
angle.

Sine θ	$\frac{O}{H}$	$\frac{3}{5}$	.6000		Cosecant θ	$\frac{H}{O}$	$\frac{5}{3}$	1.6666
Cosine θ	$\frac{A}{H}$	$\frac{4}{5}$	.8000		Secant θ	$\frac{H}{A}$	$\frac{5}{4}$	1.2500
Tangent θ	$\frac{O}{A}$	$\frac{3}{4}$	.7500		Cotangent θ	$\frac{A}{O}$	$\frac{4}{3}$	1.3333

θ is the Greek letter **theta**. It is a variable which is used to represent the
degrees of an angle when the degrees are unknown. In the function chart
above, θ is used to represent any angle. For example, the sine function can
be read: the sine of *any angle* is the hypotenuse divided into the opposite
side.

You will see the functions abbreviated quite often. Here are the standard abbreviations.

Sine θ = Sin θ	Cosecant θ = Csc θ
Cosine θ = Cos θ	Secant θ = Sec θ
Tangent θ = Tan θ	Cotangent θ = Cot θ

Practice 13: Determine the six functions for each of the angles* of these right triangles. Express your answers as decimals and round off to 4 decimal places.

Remember, there are two reference angles in each triangle.

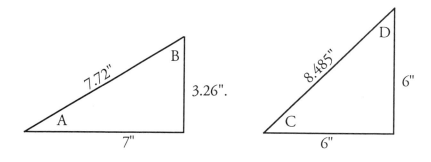

Memory Aid

There is a learning trick that can help you keep the functions straight. Simply remember this sentence: **O**scar **H**ad **A** **H**eap **O**f **A**pples. If you memorize this sentence, you can keep the functions in correct order. As you can see, this phrase uses the first letter of each of the sides.

Oscar	Had	A	Heap	Of	Apples
Opposite	Hypotenuse	Adjacent	Hypotenuse	Opposite	Adjacent

$$Sin = \frac{Opposite}{Hypotenuse} \qquad \frac{Oscar}{Had}$$

$$Cos = \frac{Adjacent}{Hypotenuse} \qquad \frac{A}{Heap}$$

$$Tan = \frac{Opposite}{Adjacent} \qquad \frac{Of}{Apples}$$

What you have to remember is sin, cos, tan, and *Oscar had a heap of apples*. The other functions are remembered by the order in which they are placed.

Sine	Cosecant
Cosine	Secant

Tangent	Cotangent

The tangent and the cotangent are easy to remember.
Look at the other two sets.

Sine	Cosecant
Cosine	Secant

You can't have the two S's or the two Co's on the same line!

The third part of remembering the order is to realize that the functions on the right side of the chart are inverse to the functions on the left side. What this means is that the numerators and denominators are reversed in position.

$$Sin = \frac{O}{H} \qquad Csc = \frac{H}{O}$$
$$Cos = \frac{A}{H} \qquad Sec = \frac{H}{A}$$
$$Tan = \frac{O}{A} \qquad Cot = \frac{A}{O}$$

If you need the secant to an angle, just remember that the secant is across from the cosine (therefore its inverse) and cosine relates to the second pair of words in *Oscar had **a heap of apples**. That means that cosine is $\frac{Adjacent}{Hypotenuse}$, so secant is $\frac{Hypotenuse}{Adjacent}$.

I hope you find this helpful.

Finding the Angles of a Right Triangle

The ratio of the sides is used in determining the degrees in an angle.

Finding the Angle Using the Calculator

The scientific calculator has a group of keys called the **arc function** keys. The arc function symbol is the $^{-1}$ in the upper right corner above the function name. *The arc function keys display the angle when given the correct function.*

There are three arc function keys on most calculators.

 Arcsine displays the angle when you enter the sine of a reference angle.

 Arccosine displays the angle when you enter the cosine.

 Arctangent displays the angle when you enter the tangent.

The three other arc functions—**arccosecant, arcsecant,** and **arccotangent** —can also be used to find the angles, but they require another step. For the present, use the arc functions that are on the calculator.

Since you are using only three arc function keys to find the angle, you need to use only three functions for your calculations. These three functions deal with all three sides of a right triangle, and Oscar Had A Heap Of Apples will help you remember these three functions.

$$\text{Sine } \theta = \frac{\text{Opposite}}{\text{Hypotenuse}}$$

$$\text{Cosine } \theta = \frac{\text{Adjacent}}{\text{Hypotenuse}}$$

$$\text{Tangent } \theta = \frac{\text{Opposite}}{\text{Adjacent}}$$

Example:

Here's how to find the degrees of the reference angle of the 3-4-5 right triangle using the calculator.

If the hypotenuse (5) is divided into the opposite side (3), the answer ($\frac{3}{5}$ = .6) is the sine of the reference angle. The $\boxed{\sin^{-1}}$ key is then used to find the degrees of the angle of that ratio.

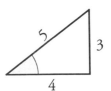

 $\boxed{\textbf{FF/Page 47}}$

Enter $3\boxed{\div}5\boxed{=}$ display $\boxed{\qquad\qquad 0.6}$

$\boxed{\sin^{-1}}$ display $\boxed{36.86989765}$

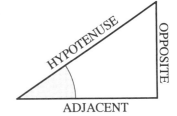

Round off to 37° (nearest $\frac{1}{2}$ °)*

As you can see, the arc function keys react to whatever is in the display. On the next practice, enter the number, look at the function to determine which arc function key to use, and then push that key.**

Practice 14: Find the angle of these functions. Round off to nearest $\frac{1}{2}$ degree.

(1) $\sin \theta = .4226$ (5) $\tan \theta = .5773$

(2) $\tan \theta = 1$ (6) $\sin \theta = .8756$

(3) $\cos \theta = .3420$ (7) $\cos \theta = .9397$

(4) $\sin \theta = 1$ (8) $\tan \theta = 2.0503$

$\boxed{\textbf{FF}}$ **Note:** The example for FF style calculator would look like this:

Push $\boxed{\sin^{-1}}$ Display $\boxed{\qquad \sin^{-1}(}$

Enter 3 $\boxed{\div}$ 5***   Display $\boxed{\quad \sin^{-1}(3\div5)}$

Push $\boxed{=}$ Display $\boxed{36.86989765}$

For (1) on Practice 14 above, the FF style calculators use this sequence:

Push $\boxed{\sin^{-1}}$ Display $\boxed{\qquad \sin^{-1}(}$

Enter .4226*** $\boxed{=}$ Display $\boxed{24.99884552}$

---

* In rounding off degrees to the nearest $\frac{1}{2}$ degree, from 0.00 to 0.249, go down to 0. From 0.25 to .749, go to .5°. From .75 to .99, go to 1°. **Also note again that this statement is incorrect for FF style calculators.
*** You may or may not need to push the right parenthesis for this calculation. It depends on the calculator.

Here is an example of finding the angles of a right triangle by using the lengths of the sides.

Follow the steps to find the angles of this right triangle.

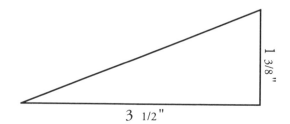

First, mark the reference angle and name the sides for that angle.

$$\text{Opposite} = 1\frac{3}{8}"$$
$$\text{Adjacent} = 3\frac{1}{2}"$$

Remember that you are working with just three functions. Look at those three and decide which one will work with the known sides.

Sine θ	$= \dfrac{\text{Opposite}}{\text{Hypotenuse}}$
Cosine θ	$= \dfrac{\text{Adjacent}}{\text{Hypotenuse}}$
Tangent θ	$= \dfrac{\text{Opposite}}{\text{Adjacent}}$

Tangent is the correct choice, because the known sides are adjacent and opposite. Tangent is the only one of the three that uses both of those sides.

$$\text{Tangent } \theta = \frac{\text{opposite}}{\text{adjacent}} = \frac{1.375}{3.5} = .392857142.$$

With that answer still in the display, push the $\boxed{\tan^{-1}}$ key*. **FF/Page 49** The display will show 21.44773633, which is the degrees of the reference angle. If we round the number off to the nearest $\frac{1}{2}$ degree, it becomes 21.5°.

---

*To find the arc tangent on some calculators, you push a $\boxed{\text{shift}}$ or $\boxed{2^{\text{nd}}}$ key, then the $\boxed{\tan}$ key. It works the same for the other arc functions as well.

Since the two angles (other than the right angle) equal a total of 90°, to find the other angle, subtract the known angle from 90°.

90° - 21.5° = 68.5°

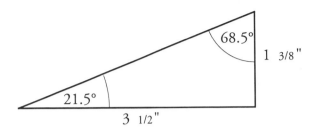

Practice 15: Find the angles of these right triangles. Round off the answer to the nearest $\frac{1}{2}$°.

(1)   Opp   22"   Hyp   34"

(2)   Adj   19.375"   Opp   27.5"

(3)   Adj   64.9375"   Hyp   78.4375"

(4)   Opp   14.5"   Adj   29.685"

(5)   Hyp   3'   Opp   2.5'

For **FF** calculators, the function must be entered before the calculation. This is true for the example on the previous page and for Practice 15. You have to look at which sides are involved, then determine which arc function is to be used. Enter that arc function, divide the sides, then push =.
Let's use the example on Page 48 to demonstrate:

Push ☐Shift☐ or ☐2nd☐ key, then ☐Tan⁻¹☐   Display

Enter 1.375÷3.5*

Push ☐=☐

Tan⁻¹(
Tan⁻¹(1.375÷3.5)
21.44773633

Notes

* You may or may not need to push the right parenthesis for this calculation. It depends on the calculator.

Finding the Angle Using the Functions Table

This is a sample of the functions table on pages 148 and 149. Notice that the table includes all six functions, plus something you will learn later, radians.

Deg↓	Radian↓	Sin θ ↓	Cos θ ↓	Tan θ ↓	Cot θ ↓	Sec θ ↓	Csc θ ↓		
36.5°	0.6370	0.5948	0.8039	0.7400	1.3514	1.2440	1.6812	0.9338	53.5°
37°	0.6458	0.6018	0.7986	0.7536	1.3270	1.2521	1.6616	0.9250	53°
37.5°	0.6545	0.6088	0.7934	0.7673	1.3032	1.2605	1.6427	0.9163	52.5°
		Cos θ ↑	Sin θ ↑	Cot θ ↑	Tan θ ↑	Csc θ ↑	Sec θ ↑	Radian↑	↑Deg

The functions table is read in two directions: from the top to the bottom and from the bottom to the top. Notice that there are angles listed on both sides. The left side of this table reads from the top to the bottom for 0° to 45° angles, and the right side reads from bottom to top for the angles 45° through 90°. When reading down the column for 0° through 45°, use the function names listed at the top of the table. When reading up the table for 45° through 90°, use the function names at the bottom. It is important that you note the difference!

We worked with the sine of an angle of the 3-4-5 right triangle earlier. Let's use the same angle for an example again. Here are the functions for a reference angle of the 3-4-5 triangle.

Sin θ	**.6000**	Csc θ	1.6666
Cos θ	.8000	Sec θ	1.2500
Tan θ	.7500	Cot θ	1.3333

Look at the functions table on pages 148 and 149.

To find the degrees of the reference angle, go down the sine column until you get to the number closest to .6000. The closest you will find is .6018. Reading the angle to the left of that number will give you the degrees of the reference angle, 37°.

Deg↓	Radian↓	Sin θ ↓	Cos θ ↓	Tan θ ↓	Cot θ ↓	Sec θ ↓	Csc θ ↓		
36.5°	0.6370	0.5948	0.8039	0.7400	1.3514	1.2440	1.6812	0.9338	53.5°
37°	0.6458	0.6018	0.7986	0.7536	1.3270	1.2521	1.6616	0.9250	53°
37.5°	0.6545	0.6088	0.7934	0.7673	1.3032	1.2605	1.6427	0.9163	52.5°
		Cos θ ↑	Sin θ ↑	Cot θ ↑	Tan θ ↑	Csc θ ↑	Sec θ ↑	Radian↑	↑Deg

Notice the degrees of the angle on the right side on the same row, 53°. This is the other angle in the 3-4-5 triangle and is called the **complementary angle**. A complementary angle is found by subtracting the reference angle from 90° (90° - 37° = 53°). Since the reference and complementary angles are in the same triangle, both angles use the same numbers (length of sides) to derive their functions. The sine of the reference angle is equal to the cosine of the complementary angle, and vice versa. The chart below shows which functions of the reference and the complementary angles are equal to each other.

Notice that all of the functions opposite each other are a name and its **co** name. That makes remembering the functions of a complementary angle easy.

Ref Angle Function	Comp Angle Function
Sine	**Co**sine
Cosine	Sine
Tangent	**Co**tangent
Cotangent	Tangent
Secant	**Co**secant
Cosecant	Secant

Note: If you name the sides wrong, you will end up with the complementary angle instead of the reference angle. One way to check yourself is to remember that the shortest side is always opposite the smallest angle.

Here is a problem which shows how to find the angles using the table.

What are the angles of this triangle?

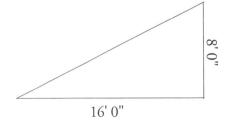

16' 0"

8' 0"

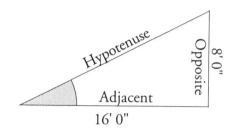

First: Draw a diagram, and label one of the angles as a reference angle. Then label the sides for that reference angle.

Adjacent = 16'   Opposite = 8'

There are two functions you can use if you know just the opposite and adjacent sides: tangent or cotangent. **Cotangent** is used this time.

$$\text{Cotangent } \theta = \frac{\text{adjacent}}{\text{opposite}}$$

$$\text{Cotangent } \theta = \frac{16}{8} \qquad \text{Cot } \theta = 2$$

Functions	
$\text{Sin } \theta = \dfrac{O}{H}$	$\text{Csc } \theta = \dfrac{H}{O}$
$\text{Cos } \theta = \dfrac{A}{H}$	$\text{Sec } \theta = \dfrac{H}{A}$
$\text{Tan } \theta = \dfrac{O}{A}$	$\text{Cot } \theta = \dfrac{A}{O}$

Look at the functions table on pages 148 and 149 under cot θ for 2. The closest angle is 26.5°, or $26\frac{1}{2}$°.

The complementary angle is 90° - 26.5° = 63.5°, or $63\frac{1}{2}$°.

Did you notice the complementary angle on the right side of the table?

Practice 16: Find the angles of these right triangles by using the functions table on pages 148 and 149.

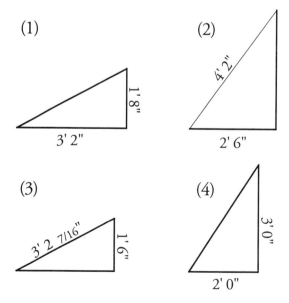

Using the Function Keys

There are three function keys on the calculator.

Sin displays the sine of an angle when the angle is entered.

Cos displays the cosine of an angle when the angle is entered.

Tan displays the tangent of an angle when the angle is entered.

FF/Page 53

Example

If the cosine of an angle is needed, the degree is entered and the cos key is pushed.

Enter 45 cos   display 0.707106781

The decimal in the display is the cosine of a 45° angle.

Practice 17: Find the indicated functions of these angles. Round off to 4 decimal places.

(1) tan 45°   (5) cos 54°

(2) sin 33°   (6) tan 75°

(3) cos 25°   (7) sin 89°

(4) sin 60°   (8) cos 60°

FF **Note:** For FF style calculator users only. If you need the cosine of a 45° angle, push the cos key, enter 45 and push the = key

Push cos Enter 45   Display   45

Push =   Display   0.707106781

The number in the display is the cosine of a 45° angle.

Using the Inverse Key

The fact that the calculator has keys for only half the functions and arc functions is not an oversight on the part of the manufacturers. It is pure economics. Why should they put six keys on a calculator when one key, the inverse key $\boxed{\frac{1}{x}}$, will do the same work?

Look at the symbol on the inverse key. It is a fraction with a variable as the denominator. That variable will be whatever is in the display when you push the inverse key. If you have 2 on the display and push $\boxed{\frac{1}{x}}$, the display will show .5, which is equal to $\frac{1}{2}$. The inverse key replaces the variable x with the number on the display and gives you the answer in decimal.

Using The Inverse Key with Functions

Let's turn to the functions and see how they work with this inverse key.

Notice that the cosecant, $\frac{H}{O}$, is the reverse of the sine, $\frac{O}{H}$. They have an inverse relationship. In fact, all the functions on the right are inverse to the functions directly across from them on the left, and vice versa.

$$\boxed{\begin{array}{ll} \text{Functions} & \\ \sin\ \theta = \dfrac{O}{H} & \text{Csc}\ \theta = \dfrac{H}{O} \\ \cos\ \theta = \dfrac{A}{H} & \text{Sec}\ \theta = \dfrac{H}{A} \\ \text{Tan}\ \theta = \dfrac{O}{A} & \text{Cot}\ \theta = \dfrac{A}{O} \end{array}}$$

A look at this chart shows these inverse relationships. To get the cosecant of an angle, first find the sine, then use the inverse key, $\boxed{\frac{1}{x}}$. If you enter the sine of an angle into the calculator and push the inverse key, you will have $\frac{1}{\sin}$, which is the cosecant of that angle.

$$\boxed{\begin{array}{ll} \sin\theta = \dfrac{1}{\csc\theta} & \csc\theta = \dfrac{1}{\sin\theta} \\ \cos\theta = \dfrac{1}{\sec\theta} & \sec\theta = \dfrac{1}{\cos\theta} \\ \tan\theta = \dfrac{1}{\cot\theta} & \cot\theta = \dfrac{1}{\tan\theta} \end{array}}$$

For example: The cosecant of 24° is found in this manner.

 **FF/Page 55**

Enter 24   Push $\boxed{\sin}$ Display $\boxed{0.406736643}$ (this is the sin of 24°)

Push $\boxed{\dfrac{1}{x}}$ Display $\boxed{2.458593336}$ (this is the cosecant)

Csc 24° = 2.4589

Practice 18: Find the functions shown. Round off to 4 decimal places.

Remember to use the calculator key that is the inverse of these functions first, then the $\boxed{\dfrac{1}{x}}$ key.

(1)   csc 39°   (5)   cot 60°

(2)   sec 45°   (6)   csc 23°

(3)   cot 30°   (7)   sec 55°

(4)   sec 15°   (8)   cot 1°

$\boxed{\text{FF}}$ **Note:** For the FF style calculator, push the appropriate function key (sin, cos, or tan), enter the angle, push =, then push the inverse key.

Using the Inverse Key with the Arc Functions

In the earlier section on arc functions, you used only the arc functions that were on the calculator. Now that you are familiar with the inverse key, you can use it to find the degrees of the angles using the other three arc functions. For example, knowing the cotangent of an angle, you can find the angle by using the inverse key, $\boxed{\frac{1}{x}}$, and the arctangent key, $\boxed{\tan^{-1}}$*.

Example: Find the angle that has a cotangent of .7002.

 $\boxed{\text{FF/Page 56}}$

Enter .7002. Push $\boxed{\frac{1}{x}}$ Display $\boxed{1.428163382}$ (tangent)

Push $\boxed{\tan^{-1}}$ Display $\boxed{55.00028982}$

55° (Rounded off to the nearest $\frac{1}{2}$°)

Practice 19: Find the angle of these functions. Round off to the nearest $\frac{1}{2}$ degree.

(1) $\cot \theta$ =   .0875   (5) $\sec \theta$ =   1.1034

(2) $\csc \theta$ =   1.0002   (6) $\cot \theta$ =   .9004

(3) $\sec \theta$ =   1.2521   (7) $\csc \theta$ =   2.9238

(4) $\cot \theta$ =   1.7321   (8) $\sec \theta$ =   5.7588

$\boxed{\text{FF}}$ **Note:** Some FF style calculators do not have an inverse key. You will have to enter the equation like this.

Push $\boxed{\tan^{-1}}$ Enter 1 ÷ 0.7002   Display $\boxed{\tan^{-1}(1/.7002)}$

Push $\boxed{=}$ Display $\boxed{55.00028982}$ angle

Notice that you are entering 1 then dividing it by the ratio 0.7002. This gives you the same answer as entering the ratio then pushing the inverse key.

---

* Remember, on your calculator you may need to use the $\boxed{\text{shift}}$ $\boxed{\tan}$ keys instead of $\boxed{\tan^{-1}}$ key.

Calculating a Right Triangle Using One Side and One Angle

Right Triangles that have the same angles also have the same ratios.

At any point along one side of an angle, a perpendicular line can be drawn to the other side and a right triangle formed. In the drawing above,

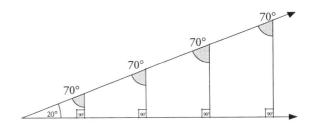

four right triangles were created this way. All four have the same angles and the same ratios between sides. The only difference is size.

If any triangles have the same angles, they also have the same side ratios. If they have the same side ratios, they have the same angles. *This is another important key in understanding the right triangle, because it allows you to calculate the length of the sides of a triangle based on the functions of a reference angle.*

> **Knowing one side and one angle (other than the right angle) allows you to calculate the measurements of the other sides and the other angle.**

Determining the unknown angle is easy; subtract the known angle from 90°, and you have the complementary angle. Finding the lengths of the other sides is a matter of rearranging the functions. When you know one angle and the length of one side, you can rearrange any of the functions to find the lengths of the other sides.

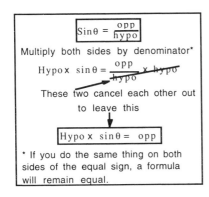

For example:

If you multiply the length of the hypotenuse times the sine of an angle, the answer will be equal to the side opposite that angle. If the adjacent side is needed, the cosine of the angle is chosen instead of the sine.

Below are a function chart and a chart that shows the rearrangement of all the functions.

$$\begin{array}{ll} \multicolumn{2}{c}{\text{Functions}} \\ \text{Sin } \theta = \dfrac{O}{H} & \text{Csc } \theta = \dfrac{H}{O} \\[6pt] \text{Cos } \theta = \dfrac{A}{H} & \text{Sec } \theta = \dfrac{H}{A} \\[6pt] \text{Tan } \theta = \dfrac{O}{A} & \text{Cot } \theta = \dfrac{A}{O} \end{array}$$

$$\begin{array}{l|l} H \times \sin \theta = O & O \times \csc \theta = H \\[4pt] H \times \cos \theta = A & A \times \sec \theta = H \\[4pt] A \times \tan \theta = O & O \times \cot \theta = A \end{array}$$

In the chart below, the terms have been placed in a different order to make the chart easier to read. Notice again that each side of a triangle can be found if either one of the other sides and an angle are known.

$$\begin{array}{l} \text{Hypotenuse} = \csc \theta \times \text{Opposite} \\ \text{Hypotenuse} = \sec \theta \times \text{Adjacent} \\ \text{Opposite} = \tan \theta \times \text{Adjacent} \\ \text{Opposite} = \sin \theta \times \text{Hypotenuse} \\ \text{Adjacent} = \cos \theta \times \text{Hypotenuse} \\ \text{Adjacent} = \cot \theta \times \text{Opposite} \end{array}$$

For example: If a right triangle has a hypotenuse of 4' and an angle of 35°, you can determine the length of the side opposite that angle by finding the formula in the chart that uses those two knowns.

Opposite = Sin θ x Hypotenuse

Opposite = sin 35° x 4'

$\boxed{\text{Sin } 35° = .573576435}$

Opposite = .573576435 x 4'

Opposite = 2.294305746'

$\boxed{\text{FF/Page 59}}$

Using the calculator:

Enter 35 $\boxed{\text{sin}}$ Display $\boxed{0.573576435}$

$\boxed{\text{x}}$ 4 $\boxed{=}$ Display $\boxed{2.294305746}$

The length of the side opposite the reference angle is 2.2943', or 2' $3\frac{9}{16}$".

There are now two ways to find the length of the adjacent side. You now know the length of two sides of the right triangle, and you know an angle. Use the side and angle method first.

To find the adjacent side when the hypotenuse and an angle are known, the cosine of the angle is used.

Adjacent = Cos θ x Hypotenuse

Adjacent = Cos 35° x 4' $\boxed{\text{Cos } 35° = .819152044}$

Adjacent = .819152044 x 4'

Adjacent = 3.276608177

 $\boxed{\text{FF/Page 59}}$
On the calculator:

Enter 35 $\boxed{\text{cos}}$ Display $\boxed{0.819152044}$

$\boxed{\text{x}}$ 4   Display $\boxed{3.276608177}$

The length of the adjacent side is 3.277', or 3' $3\frac{5}{16}$ ".

The other way of finding the adjacent side is to use two known sides.

$a = \sqrt{c^2 - b^2}$

$a = \sqrt{4^2 - 2.294305746^2}$

$a = \sqrt{16 - 5.263838854}$

$a = \sqrt{10.73616115}$

$a = 3.276608177$

Practice 20: Find the lengths of the two missing sides (the angle given is the reference angle for the side). Round off to 4 decimal places.

(1) 65°   opp   3"   (5) 27°   adj   6'
(2) 45°   hypo   3'   (6) 52°   opp   2'
(3) 10°   adj   11"   (7) 75°   hyp   13"
(4) 1°   opp   1"   (8) 87°   hyp   98'

This might be helpful on the practice above

The functions may be found with the calculator or the tables on pages 148-149.

$\boxed{\begin{array}{l} \text{Hypotenuse} = \csc θ \text{ x Opposite} \\ \text{Hypotenuse} = \sec θ \text{ x Adjacent} \\ \text{Opposite} = \tan θ \text{ x Adjacent} \\ \text{Opposite} = \sin θ \text{ x Hypotenuse} \\ \text{Adjacent} = \cos θ \text{ x Hypotenuse} \\ \text{Adjacent} = \cot θ \text{ x Opposite} \end{array}}$

$\boxed{\text{FF}}$ **Note:** For FF style calculator users, push the appropriate function button (sin, cos, or tan), enter the angle, then either push an $\boxed{=}$ $\boxed{\text{x}}$ or go straight to the $\boxed{\text{x}}$.

The 45° Right Triangle

The 45° right triangle is a special triangle, because it is the only **isosceles right triangle**, which means it has one angle of 90°, plus two sides and two angles which are equal.

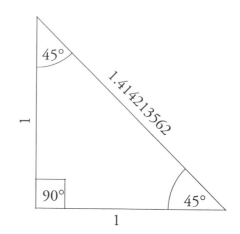

If the length of the two legs of a 45° right triangle is 1, then by using the formula for a right triangle, you will find that the hypotenuse is equal to $\sqrt{2}$.

$$a^2 + b^2 = c^2$$

$$\sqrt{a^2 + b^2} = c$$

$$\sqrt{1^2 + 1^2} = c$$

$$\sqrt{1 + 1} = c$$

$$\sqrt{2} = c$$

$$1.414213562 = c$$

Pipe fitters have a rule of thumb on 45° offsets. If the length of one of the legs is known, the length of the hypotenuse can be found by multiplying that leg times 1.4142. If the length of the hypotenuse is known and the length of the legs is needed, then multiply the hypotenuse times .7071.

Take note of how these figures are related to the functions on the next page.

Notice that since the two sides and the two angles of a 45° triangle are equal, the functions pair up. This is the reason that the pipe fitter's rule of thumb works.

$$\text{Sin } 45° = \frac{\text{opp}}{\text{hypo}} = \frac{1}{1.4142} = .7071$$

$$\text{Cos } 45° = \frac{\text{adj}}{\text{hypo}} = \frac{1}{1.4142} = .7071$$

$$\text{Tan } 45° = \frac{\text{opp}}{\text{adj}} = \frac{1}{1} = 1$$

$$\text{Csc } 45 = \frac{\text{hypo}}{\text{opp}} = \frac{1.4142}{1} = 1.4142$$

$$\text{Sec } 45° = \frac{\text{hypo}}{\text{adj}} = \frac{1.4142}{1} = 1.4142$$

$$\text{Cot } 45° = \frac{\text{adj}}{\text{opp}} = \frac{1}{1} = 1$$

The two sides, other than the hypotenuse, are called the legs of the triangle. As you can see, ***the legs of a 45° right triangle are and must be equal.*** You cannot have a 45° right triangle with unequal legs! I put a lot of emphasis on this because in the field many people seem to think otherwise, and this mistake can cause problems. Look at this drawing of a 45° simple offset. Right triangles for an offset can be drawn two different ways, and they are both correct.

Notice that the vertexes of the 45° angles are located at the center points of the elbows. That is the point that the center line of the line of pipe begins to drop or rise. The distance between the center points of the elbows is the hypotenuse of the 45° right

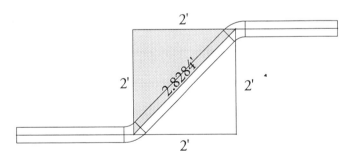

triangle. It is important to know where the triangle is located when you are doing your calculations. There is more about this in the next module.

45° Angles, Elbows, and Offsets

A 45° elbow turns a line of pipe away from the original run at a 45° angle. It is the acute angle located on the backside of the elbow that is the angle of turn or angle of rise or angle of drop. Of course, the term you use will depend on whether the pipe is rising, dropping or turning. These terms are used with any degree elbow.

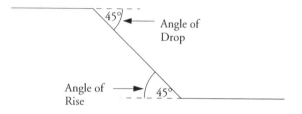

This diagram shows that the elbow turns the pipe at a 45° angle away from the straight line that the pipe had been travelling. It also shows the obtuse angle (135°) created where the center lines meet.

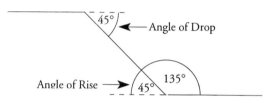

Remember that a straight line has an angle of 180°.
135° + 45° = 180°.

Notes

VII

Circles

The study of the circle is important in pipe for the simple reason that pipes are round.

Here are some terms that you will learn in this section. You might want to look them over now and turn back to them as you need the definitions.

An **arc** is a curved line.

An **arc length** is the length of the curved line.

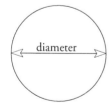 The **diameter** is a straight line from one side of a circle to the other side that passes through the center. It is abbreviated as d.

The **circumference** is the distance around a circle or a pipe. It is abbreviated as c.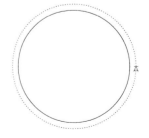

The **radius** is a straight line from the center of the circle to a point on the circle. Radius x 2 = diameter. $\dfrac{\text{Diameter}}{2}$ = radius. Radius is abbreviated as r.

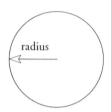

The **unit circle** is a circle with a radius of 1 unit.

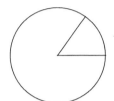

A **central angle** is an angle in which the vertex is located at the center of a circle.

Radian is a measure of arc length based on the length of the radius. One **radian** is an arc equal in length to the radius. The central angle that cuts off this arc is $\dfrac{180°}{\pi}$ or 57.295+°.

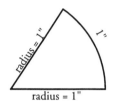

A **chord** is a straight line from one point on a circle to another point on the circle. The diameter is the longest chord of a circle.

PI π

The main symbol used when dealing with round objects is called **pi(π).** The symbol π is a letter from the Greek alphabet and is used in mathematics to represent a particular number. To a mathematician, it is the whole number 3 and an endless number of decimal places. Scientists working on space travel may use a whole page of decimal places after the number 3. The calculator key $\boxed{\pi}$ will display 3.141592654, but most of us remember π as 3.14 from our school days.

Pi is the number, that when multiplied by the **diameter** of a circle, gives you the circumference of that circle. Therefore, if

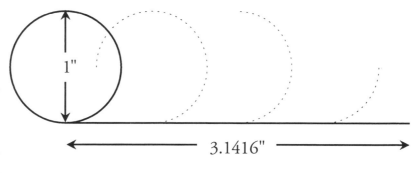

you took a 1" circle and unrolled it, it would measure approximately 3.1416".

In fact, early mathematicians rolled out all sizes of circles, and when they divided the diameter of the circle into the length rolled out, the same number, 3.14 (approximately $3\frac{1}{7}$), always appeared. The number was called pi.

Don't be confused by all the decimal places. Use whatever is in the calculator, and round the answer off to serve your need for accuracy. If you are using a calculator that doesn't have a π key, use 3.1416.

Circumference

Using the Diameter to Find the Circumference

The distance around a circle, or in our case a pipe, is called the circumference of the circle or pipe. From the section on pi, you know that the circumference of a circle is found by multiplying the diameter of the circle times pi. The formula is stated: $\boxed{c = \pi d}$

Here is an example of how to calculate circumference on your calculator.

Find the circumference of a circle that has a diameter of 6.

Push $\boxed{\pi}$ $\boxed{\text{x}}$ Display $\boxed{3.141592654}$

Enter 6 $\boxed{=}$ Display $\boxed{18.84955592}$

18.8496 is the circumference of a circle that has a diameter of 6. The unit of measurement is whichever unit of measurement you are using. It could be a unit of measurement such as inch, foot, meter, or mile. If you are using inches, then the circumference will be $18\frac{7}{8}$", but if you are using feet, it will be $18'10\frac{3}{16}$".

Practice 21: Find the circumference of these circles. Convert the decimals to fractions.

(1) d = 2'
(2) d = 3.5"
(3) d = $9\frac{7}{8}$"
(4) d = 6.75'

(5) d = 5"
(6) d = $7\frac{3}{4}$"
(7) d = 10'
(8) d = 25'

C = π d

Practice 22: Listed below are the nominal pipe size[*] and the outside diameter (O.D.) of several pipes. Use what you have learned above to find the circumference of the pipe. Use the O.D. as the d in the formula. Convert the answers to inches and fractions of an inch.

	Pipe size	O.D.
(1)	1"	1.315"
(2)	3"	3.5"
(3)	8"	8.625"
(4)	12"	12.75"
(5)	14"	14"
(6)	24"	24"

Using the Radius to Find the Circumference

The radius is $\frac{1}{2}$ of the diameter. $\boxed{\mathbf{d = 2r}}$ If the diameter is 4, then the radius is 2; if the radius is 3, then the diameter is 6. The plural of radius is radii. (In other words, we say that there is one radius but two or more radii.)

To use the radius instead of the diameter for finding the circumference of a circle, c = π**d** is stated as c = π**2r.** This formula is accurate since d = 2r.

Example: Find the circumference of a circle that has a radius of 5".

Push π x	Display 3.141592654
2 x	Display 6.283185307
5 =	Display 31.41592654

The circumference of a circle with a 5" radius is $31\frac{7}{16}$".

---

[*] Nominal pipe size is only the size that we call the pipe. The actual dimensions of the O.D. and I.D. may be different. For example: 6" pipe has an O.D. of 6.625" and an I.D. of 6.065", but the nominal pipe size is 6".

Practice 23: Find the circumference of these circles. Convert the answers to the appropriate whole units and fractions.

C = π2r

(1)   r = 1"

(5)   r = 2⅛"

(2)   r = 25'

(6)   r = 1'

(3)   r = 13"

(7)   r = 19¼"

(4)   r = 103'

(8)   r = 60'

You can now find the length of a curved line that makes one complete rotation around a center point (**circumference**). In pipe though, we deal with many curved lines (**arcs**) that only complete a fraction of a complete rotation around a center point. For example, the circumference of a circle that has a radius of 1" is 6.2832", therefore the circumference of one fourth of that circle would be equal to one fourth of 6.2832" or 1.5708". We also know from an earlier section that circle has a **central angle** of 360°. One fourth of 360° is 90°. You can combine the two statements together and say that a 90° section of a circle with a radius of 1" has an **arc length** of 1.5708". In the next sections you will learn to find the length of arcs based on knowing the radius of a circle and the central angle. This information is critical to us because it allows us to mark odd angle elbows, correctly place nozzles on tanks ands vessels, and run pipe lines around tanks parallel to the existing nozzles.

The Unit Circle

If I were to tell you that my car will travel 30 miles on 1 gallon of gas and then ask you how far I can travel on 5 gallons, you would multiply 5 times 30 and reply 150 miles. If I were to tell you that I could walk 3 miles in 1 hour and then ask how far I could walk in 3 hours, you would reply 9 miles. In each case the information I was giving you was based on the unit 1. The same type of logic has been used in the unit circle.

The unit circle is a circle with a radius of one. Because of its simplicity, the unit circle is used to show the properties of circles. If you learn the basics of the unit circle, you will then be able to apply the same principles to other circles.

Since the radius of a unit circle is always 1, replace the r in the circumference formula with 1.

$$C = 2r\pi$$

$$C = (2 \times 1 = 2)\pi$$

$$C = 2\pi \text{ or } 6.2832$$

The circumference of a unit circle is 2π, or 6.2832.

The central angle for a whole circle is 360°.

6.2832

Why is the length twice as long as the earlier diagram?

Remember that the diameter is equal to 2 radii. The diameter of a unit circle is 2.

What is the arc length of a half circle?

If a whole unit circle has a circumference of 2π, then a half circle must have an arc length of π, or 3.1416.

$\dfrac{c}{2} = \dfrac{2\pi}{2}$ which is reduced to $\dfrac{1}{2}\, c = \pi$

The central angle for a half of a circle is $\dfrac{360°}{2} = 180°$.

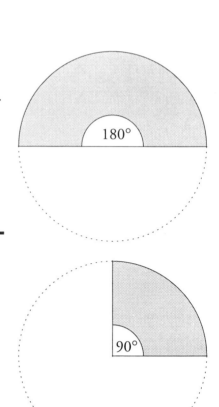

What is the arc length of $\dfrac{1}{4}$ of a unit circle?

Dividing both sides by 4 will give the answer.

$\dfrac{c}{4} = \dfrac{2\pi}{4}$ which is reduced to $\dfrac{1}{4}\, c = \dfrac{\pi}{2}$ or 1.5708

The central angle for a quarter of a circle is $\dfrac{360°}{4} = 90°$.

What is the length of the arc of $\dfrac{1}{8}$ of a circle?

Divide both sides of $c = 2\pi$ by 8.

$\dfrac{c}{8} = \dfrac{2\pi}{8}$ which is reduced to $\dfrac{1}{8}\, c = \dfrac{\pi}{4}$ or .7854

The central angle for an eighth of a circle is $\dfrac{360°}{8} = 45°$.

What is the length of the arc of $\dfrac{1}{360}$ of a circle?

The answer is reached by dividing both sides by 360.

$\dfrac{c}{360} = \dfrac{2\pi}{360}$ which is reduced to $\dfrac{1}{360}\, c = \dfrac{\pi}{180}$ or .017453292.

The central angle for $\dfrac{1}{360}$ *of a circle is* $\dfrac{360°}{360} = 1°$.

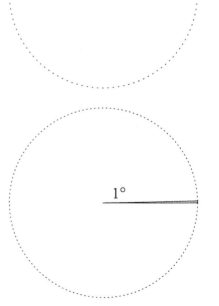

Each section on the previous page mentions a central angle and an arc length.

Here is a chart of the angles and the arc length of the unit circle.

$360° = 2\pi$	$360° = 2\pi$	$360° = 6.2832$
$\dfrac{360°}{2} = \dfrac{2\pi}{2}$	$180° = \pi$	$180° = 3.1416$
$\dfrac{360°}{4} = \dfrac{2\pi}{4}$	$90° = \dfrac{\pi}{2}$	$90° = 1.5708$
$\dfrac{360°}{8} = \dfrac{2\pi}{8}$	$45° = \dfrac{\pi}{4}$	$45° = .7854$
$\dfrac{360°}{360} = \dfrac{2\pi}{360}$	$1° = \dfrac{\pi}{180}$	$1° = .017453$

There is an established relationship between the central angle of an arc and the length of that arc. That relationship is expressed in **radians**.

Radians

The relationship between the length of an arc of any circle and its central angle is expressed in radians.

This chart shows the radians for the five central angles shown in the chart above. Notice that radian measurements are equal to the arc lengths of the unit circle. There is a more complete chart included in the functions table on pages 148 and 149.

$360°$	$= 2\pi$ radians	$= 6.2832$ radians
$180°$	$= \pi$ radians	$= 3.1416$ radians
$90°$	$= \dfrac{\pi}{2}$ radians	$= 1.5708$ radians
$45°$	$= \dfrac{\pi}{4}$ radians	$= .7854$ radians
$1°$	$= \dfrac{\pi}{180}$ radians	$= .017453$ radians

In pipe work, radians are used to find the arc lengths of curved lines. In Module Two, they are used to find the measurements of the throat and the back of odd angle elbows.

Compare the central angles, the circumferences of the unit circle (arc lengths), and radians on the chart on the next page.

Circle	Central Angle	Circumference of Unit Circle or arc length	Radians
Whole	360°	2π	2π or 6.2832
$\frac{1}{2}$	180°	π	π or 3.1416
$\frac{1}{4}$	90°	$\frac{\pi}{2}$	$\frac{\pi}{2}$ or 1.5708
$\frac{1}{8}$	45°	$\frac{\pi}{4}$	$\frac{\pi}{4}$ or .7854
$\frac{1}{16}$	$22\frac{1}{2}°$	$\frac{\pi}{8}$	$\frac{\pi}{8}$ or .3927
$\frac{1}{32}$	$11\frac{1}{4}°$	$\frac{\pi}{16}$	$\frac{\pi}{16}$ or .19635
$\frac{1}{360}$	1°	$\frac{\pi}{180}$	$\frac{\pi}{180}$ or .017453

On the right side of the chart is the column called radians. Notice that the numbers for radians are the same as the arc lengths of the unit circle.

This is one of the formulas for radians.

$$\text{Radians} = \frac{\text{Arc Length}}{\text{Radius}}$$

We will use it to show that **equal central angles have the same radians.**

#1

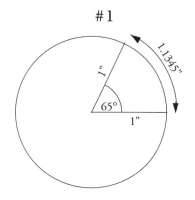

#2

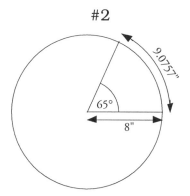

In the circles above, the 65° central angle cuts off arc lengths of 1.1345" and 9.0757". To show that the radians of those central angles are equal, divide the arc lengths by the radii of the circles.

$\theta = \dfrac{\text{arc}}{\text{radius}}$	$\theta = \dfrac{\text{arc}}{\text{radius}}$
$65° = \dfrac{1.1345}{1}$	$65° = \dfrac{9.0757}{8}$
$65° = 1.1345$ radians	$65° = 1.1345$ radians

Finding the Radians for a Central Angle

The previous example shows that equal central angles have the same radians, but how are the radians of a central angle found?

The central angle of 1° is equal to $\frac{\pi}{180}$ radians, or .017453 radians. If you multiply the central angle (65°) by the radians for 1°, the answer will equal the radians for that central angle. This is true for all central angles.

$65 \times \frac{\pi}{180} = \frac{65\pi}{180} = 1.1345$ radians.

$$\text{Angle } \theta = \theta \times \frac{\pi}{180} \text{ radians} = \frac{\theta\pi}{180} \text{ radians}$$

 Remember that θ is the variable for any angle.

Example:
Find the radians for 37°.

Angle $\theta = \frac{\theta\pi}{180}$ radians

Angle 37° $= \frac{37\pi}{180}$ radians

Angle 37° $= \frac{116.239}{180}$ radians

Angle 37° = .6458 radians

Here's how to use the calculator to find 37° in radians.

Enter 37 $\boxed{\times}$ $\boxed{\pi}$ $\boxed{=}$ Display $\boxed{116.2389282}$

$\boxed{\div}$ 180   Display $\boxed{0.645771823}$

The radians of a 37° angle are .6458.

Practice 24: Find the radians for these central angles.

(1) 10°   (5) 72°

(2) 48°   (6) 39°

(3) 57°   (7) 22°

(4) 65°   (8) 84°

Finding the Arc Length Using the Radius and Radians

If the radian measure of a 30° central angle is .5236, then the arc length of a 30° central angle with a radius of 5 is 5 x .5236, or 2.6180.

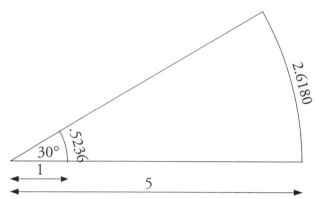

$$\text{Radius} \times \text{Radians} = \text{Arc Length}$$

Example: What is the arc length of 43° central angle located in a circle with a radius of 9"?

Arc length = radius x radians   radians = $43° \times \dfrac{\pi}{180}$ = .7505

Arc length = 9" x .7505
Arc length = 6.75" or $6\frac{3}{4}$"

Practice 25: Find the length of the arcs with these central angles and radii. Convert answers to whole numbers and fractions.

(1)	51°	6"	(7)	33°	8'
(2)	24°	12"	(8)	90°	1.75"
(3)	45°	4.5"	(9)	72°	84'
(4)	80°	3"	(10)	41°	5.465"
(5)	65°	7.5"	(11)	5°	48"
(6)	16°	11'	(12)	27°	100"

Chords

Chords are straight lines that join one point on a circle to another point on the circle.

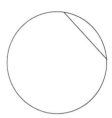

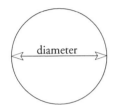

The diameter of a circle is a chord that must go through the center of the circle. It is the longest chord of each circle.

We use chords in pipe fitting to figure the distances between bolt holes when laying out flanges. Let's use flanges to learn about chords.

Flanges using the Chord Lengths

This is an eight hole blind flange.

How would you go about marking the bolt holes to be drilled? In the various pipe manuals, the charts on flanges give the diameters of the bolt hole circles and the number of holes in each particular flange. With these two pieces of information, you can calculate the distances, or lengths of chords, between the bolt holes.

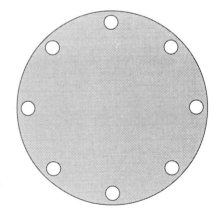

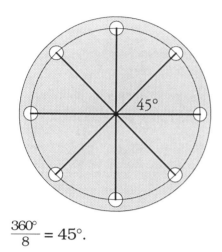

Since the holes are evenly spaced around the flange, you can find the degrees of the central angle by dividing the degrees in a whole circle by the number of holes in the flange.

An eight hole flange has a 45° central angle between the bolt holes. That angle is found by dividing 360° by 8.

$$\frac{360°}{8} = 45°.$$

When the bolt hole circle is drawn in, and a line (chord) is drawn between the two points where the angle touches the bolt hole circle, it becomes obvious that the triangle created is not a right triangle. It is an isosceles triangle.

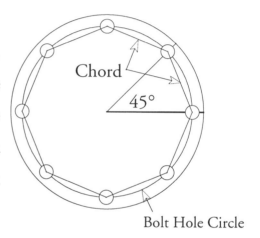

Remember, an isosceles triangle has two equal sides and two equal angles opposite those sides. There is a property of isosceles triangles that helps us calculate the length of the chord. It states that if a straight line is drawn from the *vertex* of the central angle to the center of the base of the triangle (chord), that line will divide the angle in half and be perpendicular to the base (chord). In other words, it will create two equal right triangles. Now we can use the properties of the right triangle to find the length of the chord. The length of the opposite side of each triangle is half of the length of the chord*. If you find the length of one of those sides and double it, you will have the chord length.

---

* It is half of a chord because the line was dropped from the vertex of the the angle to the *midpoint of the chord.*

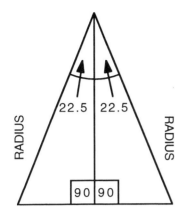

Here is the isosceles triangle without the circle. It has been divided into right triangles by a line dropped from the vertex to the midpoint.

Since the right triangles are equal, you need to work only one right triangle. The hypotenuse of the right triangle is the same as the radius of the circle. Let's work an 8" 150# flange* as an example.

An 8" 150# flange has a bolt hole circle diameter of 11.75" (radius = 5.875"). Now let's define the knowns and needs:

The **hypotenuse** is known.
The angle is **$22\frac{1}{2}$**°.

The **opposite** side is needed.

> Hypotenuse = csc θ x Opposite
> Hypotenuse = sec θ x Adjacent
> Opposite = tan θ x Adjacent
> Opposite = sin θ x Hypotenuse
> Adjacent = cos θ x Hypotenuse
> Adjacent = cot θ x Opposite

We use the chart to find the formula that will work with the information that is known.

Opposite = sin θ x Hypotenuse
Opposite = sin 22.5° x 5.875"
Opposite = .3827 x 5.875"
Opposite = 2.2484"

Chord = opposite side x 2
Chord = 2.2484 x 2
Chord = 4.496 or $4\frac{1}{2}$"

---

* Information found in the standards charts in the pipe manuals.

How does this method compare to method used in the pipe manuals?

Our formula is: They give you this way:

$Sin \dfrac{\theta}{2}$ x radius x 2= Chord $Sin \dfrac{\theta}{2}$ x diameter = chord

Sin 22.5° x **radius x 2** = chord   Sin 22.5° x diameter = chord

Since diameter = **radius x 2**

Sin 22.5° x (diameter) = chord

Now that you have seen how the chord is found, there is no reason not to use the shorter method.

Marking Flanges with a Compass

The chord length is the distance between the bolt holes, but how would you mark this distance around the bolt hole circle?

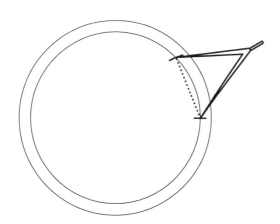

Use a compass that can maintain accuracy, set it for the chord length, and mark a starting point on the bolt hole circle. Place the sharp point of the compass on the starting mark, and mark another point on the circle. Then move to the new point and mark again. Continue marking around the circle until you arrive back at the starting mark.

If you have made correct calculations and the compass has maintained its accuracy, you should end up exactly where you started, with markings for the correct number of bolt holes.

I have made several references so far about using a compass that will maintain its accuracy. Another fitter, Gibbs Langley, and I didn't have such a compass one Saturday morning and spent quite a frustrating day trying to accurately mark

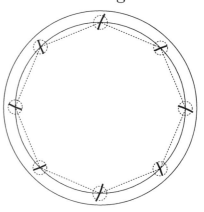

a 20" flange. The next day I made a compass, and Monday morning it took only 20 minutes to mark out the same flange. Having the right tool for the job makes a world of difference.

Note: When doing the calculations for flanges, a finer degree of accuracy is needed. Convert the decimals to fractions of $\frac{1}{32}$ of an inch.

Practice 26: Find the distance between the bolt holes for these 150# flanges.

 Remember **Chord length = Sin $\frac{\theta}{2}$ x Diameter**

	Pipe size	Bolt hole circle diameter	# of holes
(1)	4"	7.5"	8
(2)	12"	14.25"	12
(3)	16"	21.25"	16
(4)	20"	25"	20

Right Triangles in Circles

In the field, we often divide the circumference of a pipe into 8 parts, 16 parts, or 32 parts. We do this in order to lay out the pipe for the fabrication of miters, stab ins, and laterals. In the drawing below, there is an end view and a side view of a pipe with lines showing it laid out in 8 equal parts. There is also a 45° line drawn across the side view. It is a straight line, and it must be straight when you mark it on the pipe to cut a miter. On a flat surface, marking a straight line is accomplished easily with a ruler, but marking a straight line around a pipe is a little more difficult. The calculations used to accomplish these layouts are made using the right triangle and the unit circle.

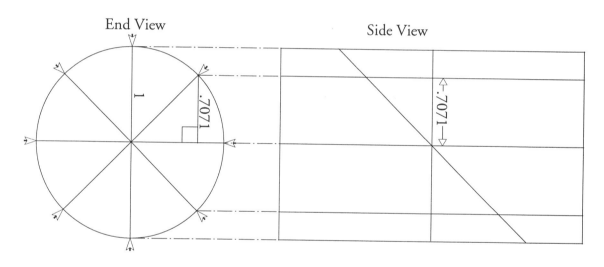

You will learn to mark out these miters in another module. Here, you will study how right triangles are calculated in the unit circle.

The triangles we draw in that circle have a hypotenuse of 1, since that is the length of the radius. We can demonstrate using just one quarter of a circle.

If the circumference of the unit circle is divided into 8 equal parts, the central angle of each part is equal to 45° ($\frac{360°}{8}$). If a line is dropped perpendicular to the base line of the quarter circle from the point where the circumference is divided, a right triangle is created.

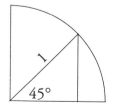

Even though this right triangle is in a circle, all the rules of the right triangle still apply. You have two knowns: the hypotenuse and one angle other than the right angle.

Hypotenuse = csc θ x Opposite
Hypotenuse = sec θ x Adjacent
Opposite = tan θ x Adjacent
Opposite = sin θ x Hypotenuse
Adjacent = cos θ x Hypotenuse
Adjacent = cot θ x Opposite

To find the length of the side opposite the central angle, multiply the length of the hypotenuse times the sine of 45°.

Opposite = Sin θ x Hypotenuse
Opposite = sin 45° x hypo
Opposite = .7071 x 1
Opposite = .7071

To find the length of the side adjacent to the central angle, multiply the length of the hypotenuse times the cosine of the central angle.

Adjacent = cos θ x hypo
Adjacent = cos 45° x hypo
Adjacent = .7071 x 1
Adjacent = .7071

If the circumference of a unit circle is divided into 16 equal parts, the central angle of each section is $22\frac{1}{2}$°. If you drop a perpendicular to the base line of the quarter circle from each point where the circumference is divided, you will create three right triangles. The central angles of these triangles will be $22\frac{1}{2}$°, 45° (which includes the central angles of two sections), and $67\frac{1}{2}$° (which includes the central angles of three sections). The opposite and adjacent sides for the 45° right triangle were worked above. To find the opposite and adjacent sides for the other two right triangles, follow the same procedures.

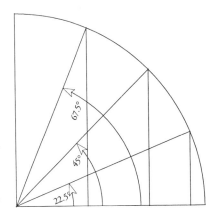

Practice 27:

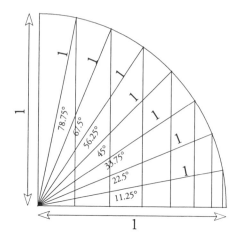

The circumference of the unit circle divided into 32 parts makes the central angle of each section $11\frac{1}{4}°$.

Below is a list of the central angles of the triangles that are found in the drawing above. Since the hypotenuse (1) and one angle are known, calculate the adjacent and opposites sides of the right triangles.

(1) $11\frac{1}{4}°$

(2) $22\frac{1}{2}°$

(3) $33\frac{3}{4}°$

(4) $45°$

(5) $56\frac{1}{4}°$

(6) $67\frac{1}{2}°$

(7) $78\frac{3}{4}°$

Look at your answers twice! There is a secret in them.

General Information

The information that you will find below is seldom used in our daily work, but you might find it useful occasionally. These formulas are within the range of what you have learned so far in this book. You will find many such formulas in the various pipe manuals, and I suggest that you work them.

Area of a Circle

The area of a circle is determined by the formula $A = \pi r^2$.

For example: The area of a 5" circle is:

π	x	r^2	= Area	
3.1416	x	2.5^2	= 19.64 in^2	

> Memory Aid: Pi are square and cakes are round!

Volume of a Cylinder

Volume = Area x length

Volume = πr^2 **x length**

For example: How much water will a 6" pipe that is 20' long hold?

The I.D. of a 6" pipe is 6.065".

The mistake that is often made with formulas is not using the same units of measure throughout the formula. All of your units of measure must be the same. Either the inches must be converted to a decimal of a foot or the feet must be converted to inches. Above you have inches and feet

If you choose **inches**, then the formula will be filled in this way.

Volume = π x r^2 x length

Volume = 3.1416 x (3.0325")^2 x 240"

Volume = 6933.679 cubic inches or 6933.679 in^3

How many gallons? There are 231 cubic inches to a gallon, so

Volume = $\dfrac{6933.679}{231}$ = 30.02 gallons

If you choose to compute the formula using **feet**, it will be worked like this:

Volume = π x r^2 x length

Volume = 3.1416 x $.2527'^2$ x 20'

Volume = 4.01228 cubic feet

How many gallons? There are .1337 cubic feet per gallon, so

Volume = $\dfrac{4.01228}{.1337}$ gallon = 30.01*

How much does it weigh? Water weighs 8.33 pounds per gallon, so 30.01 x 8.33 = 249.98 lbs.

Here's a teaser for you - How much water will a 6" 90° B.W. elbow hold? The answer is in the back along with the answers to Practice 27.

Polygons

Occasionally you will run into other geometric shapes called polygons. A polygon is a many-sided enclosed space. Rectangles and squares are examples of polygons. A **regular polygon** has equal sides and equal angles. These polygons can be inscribed in a circle, and the sides of the polygons will be chords. Since the angles are equal, you can calculate the length of the sides and the degrees of the angles from what you have learned about chords.

Example: An octagon is an eight-sided regular polygon and the distance from the center to the point where two sides meet is 5.875". How long is each of the sides? I would usually put a drawing here, but I've already drawn an octagon inscribed in a circle and calculated the length of the chords. Where? Look on page 78 in the section on 8 hole flanges! The length of the chord is the sin of 22.5 X 5.875 x 2 = 4.4965. That is the length of the sides of that regular octagon. You can figure any regular polygon in the same manner.

---

* The difference of $\dfrac{1}{100}$ of a gallon is due to the rounding off of the numbers.

Module Two

Elbows
Take Outs
Pipe Offsets

Preface

Module one covered most of the math that is used in pipe fitting. Module two applies that math to practical applications that are used in the field. The main emphasis of this module is offsets. With the math from the first module and a background in offsets, you will be able to handle most of the difficult fits required of you.

The elbow is the only fitting that is covered in this module, yet it is only one of many types of fittings used in our trade. The reason a broader range of fittings isn't included is that the odd angle elbow calculations which are shown in this module are the most difficult calculations to do. Conquer those, and the rest will be a snap.

All of the calculations in this module can be worked with your knowledge of the right triangle and the unit circle. However, it is easy to get confused if you do not solve one problem at a time. My suggestion to you is: make a thumbnail sketch of your problem, locate the triangles you need to solve, and begin by calculating the first right triangle.

In order to calculate offsets, it is first necessary for you to know how to determine take outs. In order to calculate take outs, you must first have a knowledge of the elbows. Therefore, this module will start with elbows, and continue through to combination rolling offsets.

VIII

Elbows

Three types of elbows are covered in module two: butt weld elbows, screwed elbows, and socket weld elbows. The following section is a brief description of each type.

Screwed Elbows

A **screwed elbow** is attached to the pipe by threads. The pipe, which has male threads, is screwed into the elbow, which has female threads. The elbows are factory made, with the female threads already in place. We usually cut the male threads into the pipe on the job site, using a threading machine.

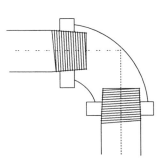

Poor threading causes the largest number of errors in the fitting of screwed pipe. Screwed piping can be fitted as precisely as any of the other types, but a great deal of care is required during the threading and assembly to do so.

Socket Weld Elbows

Socket weld elbows have a hole, called a socket, bored into the faces of the elbows. The socket is slightly larger than the O.D. of the pipe. The pipe is inserted into the socket of the fitting, pulled back $\frac{1}{16}$", then tacked. After the fit is complete, the welder finishes the filet weld.

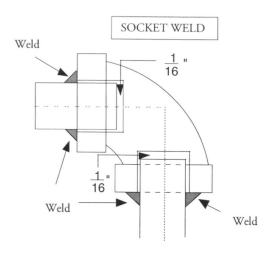

Butt Weld Elbows

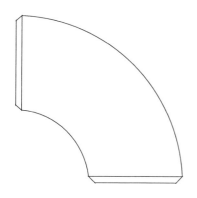

A **butt weld elbow** is not inserted into the elbow, but is held a welder's gap (see below) away from the pipe before being tacked. The pipe and the elbow are beveled, since the welder has to make a solid weld through the wall of the pipe. Unlike the other elbows, butt weld elbows can be cut and rebeveled to any angle that is needed.

Welder's Gap

In calculating fits with butt weld elbows, there is a measurement over which the pipe fitter has no control. It is called the **welder's gap**. The width of the gap is the distance that the welder feels comfortable with, and what he feels will help him make the best weld. It is best to ask the welder the gap width he wants before making any calculations.

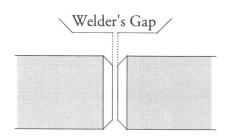

Marking Butt Weld Elbows

In the field, it is often required that 90° butt weld elbows be cut down to smaller degree elbows. If these elbows are marked and cut correctly, not only will the angle of the elbow be correct, but the face of the elbow will remain circular instead of becoming egg shaped. Welders do not take too kindly to egg shaped fits, so try to avoid them in order to keep peace in the family.

In the first module, you learned to calculate arc lengths using the radians of a central angle and a radius. You are now going to put that knowledge to use in learning how to mark elbows.

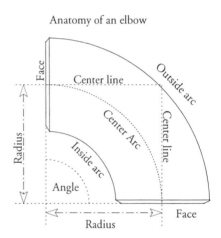

Anatomy of an elbow

When there is a need to cut a smaller degree elbow, there are two things that need to be accomplished with your cut. One: you must cut the elbow to the exact angle needed. Two: you must cut the elbow in such a way that the opening at the face will remain round. The opening must remain circular in order to get a good fit for the weld. To accomplish that, the radius for the cut elbow must remain the same as the radius of the original elbow.

To cut an elbow down to a smaller degree, you need a line as a guide by which to cut. In order to create a line, you need at least two points to connect. There are two places where we can calculate the correct angle for the new elbow using the radius. They are the **inside arc of an elbow** and the **outside arc of an elbow**, also known as the **throat** and **back of the elbow.** Since the throat and the back of the elbow are arcs, the formula for arc length is used to determine the distance from the face of the elbow to a point on the line of cut.

To find the two arc lengths needed, use the formula for arc length. As you can see, you need to know the radians for the degree elbow, and the radius.

Arc Length = Radians x Radius

The first calculation is to find the number of radians in the degree of the angle of the new elbow.

The first calculation is to find the number of radians in the degree of the angle of the new elbow.

$$\text{Angle } \theta = \frac{\theta \times \pi}{180} \text{ radians}$$

Remember, equal angles have the same radians.

The second part of the formula is **radius**. For each elbow, there are three radii: the center radius, the inside radius, and the outside radius. While we need only the dimensions of the inside and outside radii for our calculation, the center radius is needed to find them.

We refer to the **center radius** of an elbow as the **radius of the elbow.** *The center radius for B.W elbows is $1\frac{1}{2}$ times the nominal pipe size*. This is one of your knowns, since you always know the pipe size with which you are working. (Two exceptions are the $\frac{1}{2}$" and $\frac{3}{4}$" butt weld elbow which has a $1\frac{1}{2}$" radius.)

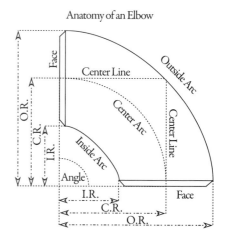

Anatomy of an Elbow

The **inside radius of the elbow** is found by subtracting $\frac{1}{2}$ the O.D. of the pipe from the center radius of the elbow.

The **outside radius of the elbow** is found by adding $\frac{1}{2}$ the O.D. of the pipe to the center radius of the elbow.

Look at this drawing of a 6" B.W. elbow to see how the lengths of the radii are found.

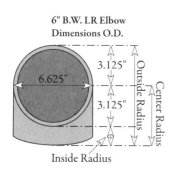

6" B.W. LR Elbow
Dimensions O.D.

The center radius
6" x 1.5 = 9".
The O.D. is 6.625"
$\frac{1}{2}$ O.D. is 3.3125".
The inside radius
9" - 3.3125" = **5.6875"**.
The outside radius
9" + 3.3125" = **12.3125"**

---

*Nominal pipe size (NPS) is the size we call the pipe, not to be confused with the actual size of the pipe. A 6" pipe has an inside diameter (I.D.) of 6.065" and an outside diameter (O.D.) of 6.625". **6" is the nominal pipe size.**

You now have all the information needed to calculate the two arc lengths using the arc length formula.

The *outside* arc length is found by multiplying the *outside* radius times the radians of the central angle of the elbow.

The *inside* arc length is found by multiplying the *inside* radius times the radians of the central angle of the elbow.

Example: Find the measurements for cutting a 6" 37° B.W. elbow.

First	Find the radians for 37°.	**Angle** $\theta = \dfrac{\theta\pi}{180}$ **radians**[*] $37° = \dfrac{37 \times 3.1416}{180}$ radians $37° = .6458$ radians
Second	Find the inside and outside radii of a 6" 90° B.W. elbow.	The inside radius is 5.6875". The outside radius is 12.3125".
Third	Find the inside and outside arc lengths of the 37° elbow.	**radians x radius = arc length** The inside arc: $.6458 \times 5.6875" = 3.6730"$ or $3\frac{11}{16}"$ The outside arc: $.6458 \times 12.3125" = 7.9511"$ or $7\frac{15}{16}"$

These are the actual measurements used to mark a 6" 37° B.W. elbow. There is a **center arc** that is seldom calculated, because the sides of the elbow are usually marked using a square and a protractor.

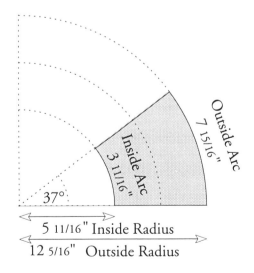

---

[*] Some people prefer to memorize .017453, which is $\dfrac{\pi}{180}$, and then multiply θ x .017453.

Practice 28: Find the inside and outside arcs for B.W. elbows for the angle and pipe size given.

$$\boxed{\text{Arc Length = Radians x Radius}}$$

	Angle	NPS	Pipe O.D.		Angle	NPS	Pipe O.D.
(1)	60°	8"	8.625	(5)	30°	12"	12.75"
(2)	45°	4"	4.5	(6)	55°	18"	18"
(3)	67°	2"	2.375	(7)	15°	6"	6.625"
(4)	22°	10"	10.75	(8)	$86\frac{1}{2}$°	14"	14"

There is a chart in the back of the book that shows the inside and outside radii of the different size elbows. However, it is best to learn the process for finding the two radii first, then use the chart with the problems in later sections to save time.

Easy Way Out

The procedures shown on the previous pages allow you to calculate the arc lengths of the elbows without physically having an elbow. There is another method that works when you are in the field with the actual elbow: Measure the back of the elbow, divide the distance by 90, then multiply the answer times the degree of the elbow you need. The answer is the measurement for the back of the new elbow. Repeat the procedure for the throat.

What you have done is found the arc length for one degree, then multiplied it times the desired angle. It is important that you measure both of the new arc lengths from the same face.

Marking a 45° B.W. Elbow

The elbow that is cut most often in the field is the 45° B.W. elbow. You can use the methods that you have just learned; however, *the trick of the trade* method is quicker and simpler. The logic of this method is simple: Since 45° is $\frac{90°}{2}$, a 90° elbow cut in half will yield two 45° elbows.

Measure the back of the elbow, and divide the distance by 2. That is the arc length for the 45° elbow. Use that measurement to mark the back of the elbow.

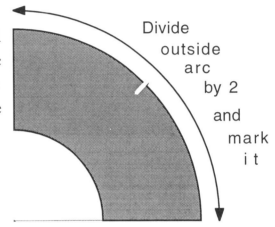

Divide outside arc by 2 and mark it

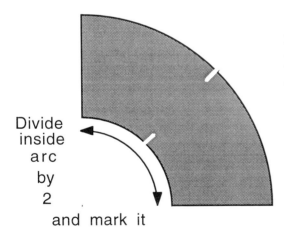

Divide inside arc by 2 and mark it

Measure the throat, and divide that measurement by 2. That is the inside arc length for the 45° elbow. Measure up the throat and mark the new length on it. Remember, it is important to measure from the same face for both the throat and the back of the elbow. It doesn't make any difference with 45° elbows, but it will with other elbows, so develop the habit of doing it right.

Marking the sides of the elbows can be difficult, except in this case. Find a scrap piece of wood or anything that has parallel sides about half the height of the elbow. Use it as a marking block.

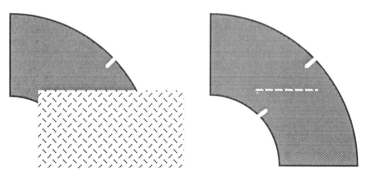

With the elbow and the marking block on a flat surface, place the block beside the elbow and mark a line along the top of the block on the side of the elbow. Turn the elbow over and set it on the other face, keeping the marked side toward you. Place the block beside the elbow again and mark another line across the side.

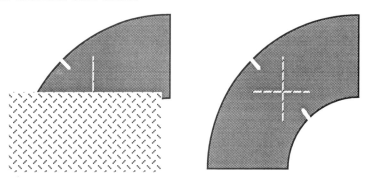

Repeat the process on the other side of the elbow.

The point at which the lines cross will probably not be the exact center of the side of the elbow; however, it will be a point on the cutting line.

You now have four marks on the elbow. Connect the four marks with a soapstone. (For the smaller elbows, I use a small pocket rule that has a $\frac{1}{4}$ " wide blade as a guide. For the medium size elbows, I use a broken portaband blade, and for the larger elbows, a wraparound.) When you look at the elbow from the side, the mark should be a straight line.

IX

Take Out

Every line of pipe that is run involves the calculations of **take outs**. As you can see in the drawing, the center line is extended a certain distance beyond the pipe by the the elbow.

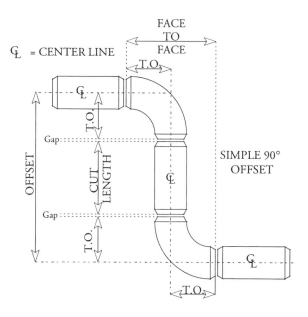

The take out *of a fitting is the distance that a fitting extends the center line of a run of pipe past the end of the pipe.* In order to find the correct length to cut a pipe for an offset, it is necessary to subtract the take outs and welder's gaps from the overall length of the run. For that reason, determining the correct take out for the fittings being used is essential to good pipe fitting.

Note on the drawing that the face to face measurement is equal to two take outs. This is true for all 90° offsets.

Take Out Formula

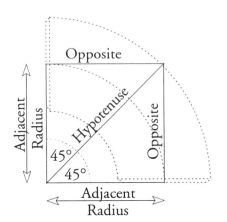

One of the main goals of this book is to show how to locate the right triangles in the problems. They are often hard to spot.

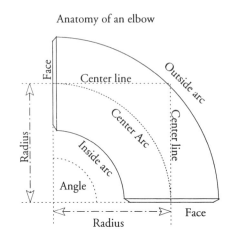

Anatomy of an elbow

The take out formula is a good example. The take out formula is derived from a pair of right triangles that are in every elbow. The adjacent legs of these right triangles are formed by lines that come from the vertex of the angle of the elbow, to the center of the faces. The center lines that come into the elbow from the faces form the opposite legs. The hypotenuse is created by a line that divides the angle in half and ends at the point where the center lines of the elbows meet.

The drawing shows the two triangles being discussed. Notice that the adjacent sides are equal to the radius. (The radius of a long radius elbow is $1\frac{1}{2}$ times the NPS except for the $\frac{1}{2}$" elbow). *The opposite sides are equal to the take out for the elbow.* You can also see that the angle located at the vertex is one half of the angle of the elbow.

State the knowns and needs:

Known: The angle

(always half the angle of the elbow)

The adjacent side

(the radius, which is $1\frac{1}{2}$ x NPS except for the $\frac{1}{2}$" elbow)

Need: The opposite side

(the take out of the elbow)

Hypotenuse	= csc θ x Opposite
Hypotenuse	= sec θ x Adjacent
Opposite	= tan θ x Adjacent
Opposite	= sin θ x Hypotenuse
Adjacent	= cos θ x Hypotenuse
Adjacent	= cot θ x Opposite

The chart shows that the formula needed is:

Opposite = tan θ x adjacent

Each time you calculate a take out, you could go through the right triangle and this formula, but let's look at this formula in a different way. We already know that the opposite side is the take out of the elbow, so you can substitute take out for opposite:

Take out of the elbow = Tan θ x adjacent.

We also know that the adjacent side is equal to the radius of the elbow:

Take out of the elbow = Tan θ x radius of the elbow.

We know, too, that the angle of the triangle is equal to $\frac{1}{2}$ the angle of the elbow:

Take out of the elbow = Tan $\frac{θ}{2}$ of the elbow x radius of the elbow.

The take out formula is shortened to:

$$\text{Take out} = \text{Tan}\frac{θ}{2} \text{ x radius of elbow}$$

This formula works for all B. W. elbows. Below are drawings showing the triangles within a 45° elbow and a 37° elbow.

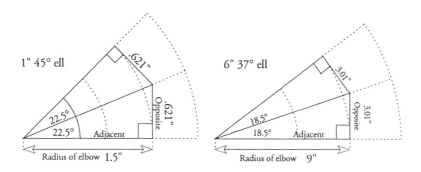

As you can see, the process above would be used to find the opposite sides for these triangles. Therefore, the take out formula works for these elbows as well.

Butt Weld Elbows

90° B.W. Elbows

Look at this drawing of a 1" 90° long radius B.W. elbow. The radius of the elbow is shown by the dotted lines on the bottom and on the left. The take out is shown by the dotted lines on the right and the top. The radius and take out for 90° B.W. elbows are equal.

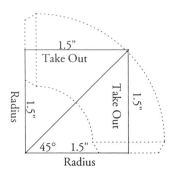

To show why they are equal, let's use the formula for take out.

$$\text{Take out} = \tan\frac{\theta}{2} \times \text{radius of elbow}$$

Take out $= \tan\dfrac{90°}{2} \times$ radius of elbow   Radius $= 1\frac{1}{2}"$

Take out $= \tan 45° \times 1\frac{1}{2}"$ Tan $45° = 1$

Take out $= 1 \times 1\frac{1}{2}"$

Take out $= 1\frac{1}{2}"$

*On **90° elbows only**, the radius and the take out are always equal, because the tangent for 45° is 1. The radius for elbows cut from a 90° elbow will remain the same as the 90° elbow, but the take out will change depending on the degree of the elbow.*

Do not use $\dfrac{\tan\theta}{2}$ as a substitute for $\tan\dfrac{\theta}{2}$. In other words, don't take the tangent of the whole angle and divide it by 2. Divide the angle by 2, and then get the tangent of the half angle. It makes a difference.

45° B.W. Elbow

The 45° elbow is the next most commonly used B.W. elbow. We have to split the discussion of the take outs for 45° B.W. elbows into two sections: the field cut 45's, and the factory made 45's.

Field Cut 45° B.W. Elbows

If you cut a 90° B.W. long radius elbow down to a 45° elbow, the take out formula will give you the correct take out.

Example: Here is how you would use the take out formula to calculate a 1" **field cut** 45° B.W. elbow.

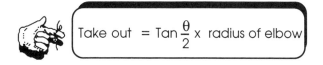

$$\text{Take out} = \text{Tan}\, \frac{\theta}{2} \times \text{radius of elbow}$$

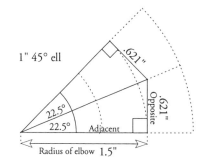

1" 45° ell

Take out $= \tan \dfrac{45°}{2} \times 1.5"$

Take out $= \tan 22.5° \times 1.5"$ (tan 22.5° = .4142)

Take out $= .4142 \times 1.5"$

Take out $= .621"$

Factory Made 45° B.W. Elbows

It is necessary to separate the factory made 45° B.W. elbows from the field cut elbows for two reasons: First, all the factory made 45° elbows below 4" have longer than standard radii; therefore, they have different take outs. Second, if the take out formula is used with factory made elbows (10" and above), the answers are $\frac{1}{16}$" off the standards chart.

The 45° B.W. elbow Take Outs		
Pipe size	Field Cut	Factory Made
½"	⅝"	⅝"
¾"	⅝"	¾"
1"	⅝"	⅞"
1 ½"	¹⁵⁄₁₆"	1 ⅛"
2"	1 ¼"	1 ⅜"
2 ½"	1⁹⁄₁₆"	1 ¾"
3"	1 ⅞"	2"
4"	2 ½"	2 ½"
5"	3 ⅛"	3 ⅛"
6"	3 ¾"	3 ¾"

The 45° B.W. elbow Take Outs		
Pipe Size	Field Cut	Factory Made
8"	5"	5"
10"	6 ³⁄₁₆"	6 ¼"
12"	7 ⁷⁄₁₆"	7 ½"
14"	8 ¹¹⁄₁₆"	8 ¾"
16"	9 ¹⁵⁄₁₆"	10"
18"	11 ³⁄₁₆"	11 ¼"
20"	12 ⁷⁄₁₆"	12 ½"
22"	13 ¹¹⁄₁₆"	13 ½"
24"	14 ¹⁵⁄₁₆"	15"

Differences in factory made and field cut 45° B.W. Elbows.

Below 4"	The difference in the take outs varies from $\frac{1}{32}$" to $\frac{1}{4}$"
From 4" to 8"	Both types have the same take out.
10" and up	There is a slight difference between the standards set for the manufacturers and the elbows we cut in the field.

The differences between the two types may seem slight. However, in off-sets, elbows are used in pairs. Therefore, the difference is doubled.

If you use factory made 45° B.W. elbows, use the standards charts in your pipe manuals for the take out.

If you use field cut 45° elbows, use the take out formula.

---

[*]This standard should be 13 ¾". However, until the standard is changed, 13 ½" will remain as the standard.

Odd Angle Elbows

An **odd angle elbow** is any elbow other than a 90° or 45°. When using odd angle B.W. elbows in an offset, the take out is found by using the take out formula.

$$\text{Take out} = \text{Tan } \frac{\theta}{2} \text{ x radius of elbow}$$

Here's how to calculate the take out for a 6" 37° B.W. elbow.

Take out = $\tan \dfrac{37°}{2}$ x radius of elbow

Take out =   tan 18.5° x 9"

Take out =   .3346 x 9"

Take out =   3.011"

Radius of 6" elbow = 9"

tan 18.5° = .3346

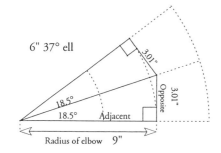

6" 37° ell

3.01"

18.5°

18.5°   Adjacent

3.01" Opposite

Radius of elbow   9"

Practice 29: Find the take out for these B.W. elbows.

	Pipe size	Degree		Pipe size	Degree
(1)	12"	28°	(6)	1"	85°
(2)	8"	70°	(7)	10"	60°
(3)	6"	33.5°	(8)	4"	48°
(4)	18"	30°	(9)	24"	10°
(5)	2"	20°	(10)	14"	51°

Screwed and Socket Weld Take Outs

Remember that take out is the distance that the fitting extends the center line of the pipe past the end of the pipe.

There are several kinds of screwed and socket weld elbows; however, I tend to approach them all in the same manner. I measure the elbow from its center to end, then subtract the distance that the pipe is inserted into the elbow. The answer is the take out. The amount of pipe that is inserted into the elbow is called the **make up**.

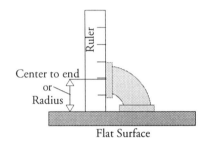

90° Screwed Elbows

Look at this drawing of a 1" 90° screwed elbow.

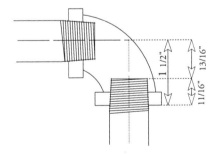

The distance from the center to end of the elbow is $1\frac{1}{2}$". The make up of the threads is $\frac{11}{16}$".

$$1\frac{1}{2}" - \frac{11}{16}" = \frac{13}{16}"$$

The take out* is $\frac{13}{16}$".

---

* The take out will not be correct if the threads are cut too deep or too shallow or tightened too much. The key to the correct take out for a screwed fitting is not just in the calculations, but also in the threading and tightening of the pipe. In the field, with a number of people using the same machine, it can be difficult to maintain the correct thread depth.

90° Socket Weld Elbows

With socket weld elbows, the pipe is pushed all the way into the socket, then pulled back $\frac{1}{16th}$ of an inch before the first tack is made. The $\frac{1}{16}$" must be taken into consideration when calculating the take out of socket weld fittings.

The socket in this drawing is $\frac{11}{16}$" deep, but the make up is $\frac{1}{16}$" less.

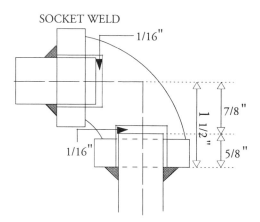

SOCKET WELD

1/16"

1/16"

7/8"

1 1/2"

5/8"

45° Screwed and Socket Weld Elbows

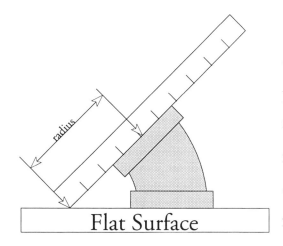

radius

Flat Surface

To find the take out for any type of 45° elbow, put the fitting on a flat surface and place a ruler on the face of the elbow. Slide the ruler down until it touches the flat surface, and read the rule at the center of the face of the elbow. That measurement will be the radius of the elbow. Use this radius in the formula for take outs to calculate the center to end distance for the elbow. With socket weld and screw elbows, once you have found the center to end measurement, subtract the *make up* from the center to end measurement to find the actual take out.

Notes

X

Pipe Offsets

In piping, to **offset** *means to continue a line of pipe on a new elevation, or in a new direction, or both.* An offset of pipe is used to make a transition from an original elevation or direction to a new elevation or direction.

Most offsets are simple, involving just 90° elbows, but at times there are other lines in the way that cause the offset to be more complex. Whether the offsets are simple or complex, they can all be calculated using the properties of the right triangle.

When calculating offsets, remember that they are calculated with the center lines of the pipe. The size of the pipe and the type of the elbow do not enter into the calculations until it is time to find the cut length of the pipe.

One of the tricks of the trade is to draw the offset in a box. The box has 6 sides that are rectangular planes, and there are several other planes that can be formed on the inside of the box. If the sides are divided by drawing a diagonal line between opposite corners, you create two right triangles. Here is a sample box.

Offset Box

As you can see, the inside of the box has more rectangles and triangles. Since you can calculate right triangles, you should be able to figure any angle and the length of any line in the box. Work one triangle at a time to keep from becoming confused.

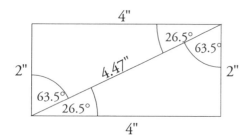

It is important to realize that if a line divides a rectangle into two parts by connecting opposite corners, it will create two right triangles that are equal in every way.

The diagonal line is a common side of the two triangles.

26.5° and 63.5° are complementary angles. If they are added together, their sum is 90°, which is the angle of the corner of the rectangle.

As you can see in the drawing, the lengths of the sides opposite each other are equal. If all four sides were equal to each other, the rectangle would be a square, and the right triangles would be 45° right triangles.

Practice 30: Sketch the following rectangles. Divide them into triangles with a diagonal line, then find the angles and the length of the diagonal. The dimensions in each problem are the two legs of a right triangle. Round off the angles to the nearest half degree.

(1)	6"	13"		(7)	1'6"	3'9"
(2)	47"	33"		(8)	3'	3'
(3)	14"	19"		(9)	57"	21"
(4)	2'	3'1"		(10)	4'3"	6'8"
(5)	$2\frac{9}{16}$"	$4\frac{3}{8}$"		(11)	12.5"	13.75"
(6)	5"	9"		(12)	23"	23"

All of the examples in this section were for butt weld elbows; however, you should be able to make the conversion for other types of elbows, since the main difference is the take out of the elbows and the cut lengths of the pipe. You can do simple and rolling offsets with screwed and socket weld fittings. The only offsets you can't do with screwed or socket weld fittings are those with odd angle elbows.

Simple Offsets

When you are running a line that is level to the ground, a **simple offset** will move the center line of the pipe to another position that is either level or plumb with the starting center line. With a line that is plumb to the ground, the simple offset will move the center line either north, south, east, or west.

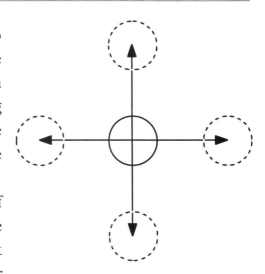

The reason that it will move in one of these main directions is that most of the equipment and structures on our jobs are set to major lines of the compass. Since our measurements usually come from the beams or columns, the measurements for offsets will come from the center lines of those beams and columns.

90° Simple Offset

The most common offset is a 90° simple offset. It moves a line of pipe to a new location using 90° ells*. A simple 90° offset requires little calculation. The length to cut the pipe is usually the most important consideration. Here is a 1" butt weld line with a 90° simple offset.

The offset is $5\frac{1}{2}$". To find the cut length of the pipe, the take out of two 90° elbows and two welder's gaps are subtracted from the length of the **run**. *The run is the path that the pipe takes to get to the new center line.* The take out and the welder's gap have been discussed previously. My welders generally ask for a $\frac{3}{32}$" gap, but $\frac{1}{8}$" is also commonly used.

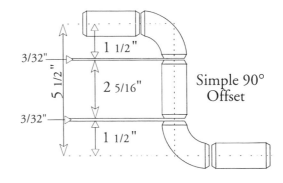

Simple 90° Offset

---

* Ell is a shortened name for elbow.

Run - 2 T.O. - 2 gaps = cut length

5.5" - 2(1.5") - 2(.09375") = cut length

5.5" - 3" - .1875" = 2.3125" or $2\frac{5}{16}$"

Practice 31: Find the cut length of the 90° offsets. Use a $\frac{1}{8}$" gap, and remember that the take out for a 90° elbow is 1.5 times NPS.

	offset	pipe size		offset	pipe size
(1)	24"	6"	(5)	17"	3"
(2)	6'10"	2"	(6)	7'7"	1"
(3)	11"	1.5"	(7)	19'11"	8"
(4)	1'2"	3"	(8)	$22'4\frac{1}{2}$"	12"

45° Simple Offset

The 45° simple offset is the next most common offset. This offset again will stay level or plumb with the original center line, but will travel to the new center line at a 45° angle.

6"std wt butt weld

45° simple offset

24"

24"

31.5"

Notice the triangle created with the dotted lines. It is a 45° right triangle, with legs 24" in length. The vertexes of both 45° angles of the right triangle are located at the center of an elbow. In order to fabricate the offset, you need to know the length of the hypotenuse of the triangle.

To find the hypotenuse of the triangle, you can use either the cosecant and the opposite side, or the secant and the adjacent side.* For this example, we will use the cosecant and opposite.

Hypotenuse = csc θ x Opposite
Hypotenuse = sec θ x Adjacent
Opposite = tan θ x Adjacent
Opposite = sin θ x Hypotenuse
Adjacent = cos θ x Hypotenuse
Adjacent = cot θ x Opposite

Hypotenuse = Csc θ x opposite

Hypotenuse = Csc 45° x 24"   Csc 45° = 1.4142

Hypotenuse = 1.4142 x 24"

Hypotenuse = 33.9408"

---

* You can use $\sqrt{a^2+b^2}=c$.

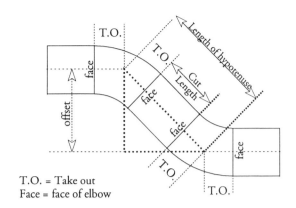

T.O. = Take out
Face = face of elbow

To determine the cut length of the pipe, you need to subtract two 45° take outs and 2 welder's gaps from the length of the hypotenuse.[*]

The take out for a 6" 45° B.W. elbow is 3.75".

Run - 2 T.O. - 2 gaps = **cut length**

33.94" - 7.5" - .1875" = **26.2525"**

The cut length for the pipe is $26\frac{1}{4}$".

Practice 32: Find the cut lengths for these 45° offsets (the welder's gap is $\frac{3}{32}$"). Use field cut elbows.

Remember that both sides of a 45° right triangle are equal.

	Offset	NPS[**]		Offset	NPS
(1)	37"	2"	(7)	27"	12"
(2)	14"	3"	(8)	41"	6"
(3)	63"	14"	(9)	56"	4"
(4)	4'	1"	(10)	5"	$\frac{1}{2}$"
(5)	7'7"	8"	(11)	21'	20"
(6)	11'4"	$1\frac{1}{2}$"	(12)	$18'5\frac{3}{4}$"	10"

---

[*] When the hypotenuse is also the center line of the pipe it is called the run.

[**] Nominal pipe size.

Face to Face

Every offset has a **face to face** measurement. *It is the distance between the parallel faces of the elbows of an offset.* The length of the face to face for this 45° simple offset is figured by *adding the length of the bottom leg and two take outs for 45° elbows.* In the sketch of the previous 6" pipe example, the face to face measurement is 24" + 7.5",

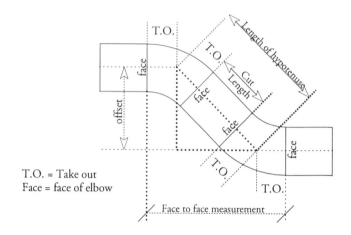

T.O. = Take out
Face = face of elbow

making a total of 31.5", or $31\frac{1}{2}$". The face to face measurement of a 24" 45° offset of 6" B.W. pipe *can only be* 31.5". I emphasize this because some people have a tendency to think that this number is changeable. It is changeable only if you change the length of the offset, the angle of the elbows, or the size of pipe.

The face to face of the other types of offsets are not covered here, but you will benefit by calculating some of them on your own.

Practice 33: Find the face to face measurements of each of the offsets in Practice 32.

The Shortest 45° Simple Offset

The shortest 45° simple offset is two 45° elbows welded together. How much would that offset the center line of the pipe? Using 4" as the pipe size, let's use the knowns to calculate the length of the offset. The knowns are the take outs and radius for the 45° elbow. Since there is no

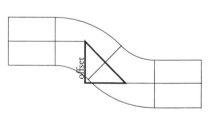

pipe involved in the run of the offset, the hypotenuse of the 45° right triangle is just two take outs and one gap. The take out for each 4" factory made 45° elbow is 2.5", and one gap is .09375"; therefore, the length of the hypotenuse is 5.09375".

To find the length of the offset, find the length of the opposite.

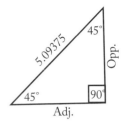

Opposite = sin θ x hypotenuse

Opposite = sin 45° x hypotenuse

Opposite = 0.7071 x 5.09375"

Opposite = 3.6018 or 3 ⅝"

This is a chart which shows the shortest 45° offsets. Notice that the chart is divided into two sections: field cut 45° ells and factory made 45° ells. A quick look at the numbers show why they must be shown separately.

Pipe Size	Take out Field Cut 45's	Offset Field Cut 45's	Take out Factory Made 45's	Offset Factory Made 45's
½"	0.625"	0.9502"	0.625"	0.9502"
¾"	0.625"	0.9502"	0.75"	1.127"
1"	0.6213"	0.9450"	0.875"	1.3037"
1¼"	0.7767"	1.1646"	1"	1.4805"
1½"	0.9320"	1.3843"	1.125"	1.6573"
2"	1.2426"	1.8236"	1.375"	2.0108"
2½"	1.5533"	2.2630"	1.75"	2.5411"
3"	1.8640"	2.7023"	2"	2.8947"
3½"	2.1746"	3.1416"	2.25"	3.2482"
4"	2.4853"	3.5810"	2.5"	3.6018"
5"	3.1066"	4.4596"	3.125"	4.4857"
6"	3.7279"	5.3383"	3.75"	5.3695"
8"	4.9706"	7.0957"	5"	7.1373"
10"	6.2132"	8.8530"	6.25"	8.9050"
12"	7.4558"	10.6103"	7.5"	10.6728"
14"	8.6985"	12.3677"	8.75"	12.4405"
16"	9.9411"	14.1250"	10"	14.2083"
18"	11.1838"	15.8824"	11.25"	15.9760"
20"	12.4264"	17.6397"	12.5"	17.7438"
22"	13.6690"	19.3971"	13.5"	19.1580"
24"	14.9117"	21.1544"	15"	21.2793"

Odd Angle Simple Offsets

Odd angle simple offsets are used when a 45° or 90° offset will not work. They require more work, since the elbows have to be calculated and cut; however, sometimes you don't have any choice but to use them. In the previous offsets, the angles of the elbows were known. In this offset, you know just the lengths of the legs. *Odd angle simple offsets always have legs that are unequal in length.*[*]

To solve odd angle offsets, I usually find the angle of the elbows, the length of the run, and then the cut length.

To find the angle of the elbows, you must find the angle of rise or angle of drop.

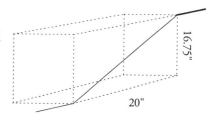

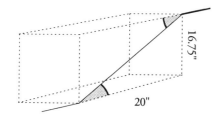

As you can see in the drawing, the angle of rise and the angle of drop for simple offsets are located at the backs of the elbows. With the position of the reference angle known, you can name the sides and use the arc function to determine the degree of the angle.

20" is the adjacent side for that reference angle and $16\frac{3}{4}$" is the opposite side.

$$\text{Tangent} = \frac{\text{opposite}}{\text{adjacent}} = \frac{16\frac{3}{4}"}{20"} = .8375 \boxed{\tan^{-1}} = 39.95°$$

Both elbows have an angle of 40°.

If this is a 14" line of pipe, then the cut length of the pipe is figured in this manner:

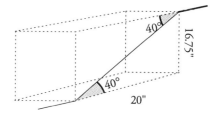

---

[*] The 45° simple offset has equal length legs.

First, you would need the length of the hypotenuse.

$$a^2 + b^2 = c^2$$
$$\sqrt{a^2 + b^2} = c$$
$$\sqrt{16.75^2 + 20^2} = c$$
$$\sqrt{280.5625 + 400} = c$$
$$\sqrt{680.5625} = c$$
$$26.0875 = c$$

The length of the hypotenuse, or run, is 26.0875".

Second, you need the length of the take out for a 14" 40° elbow. That is found using the take out formula.

Take out = tan $\frac{\theta}{2}$ x radius   radius of 14" ell = 21"

Take out = tan 20° x 21"   tan 20° = .3640

Take out = .3640 x 21"

Take out = 7.6434

Third, you need the cut length of the pipe.
The formula for cut length is

Cut length = Run - 2 T.O. - 2 Gaps   Gap is $\frac{1}{8}$"

Cut length = 26.0875 - 2(7.6434) -2(.125)

Cut length = 26.0875 -15.2868 - .25

Cut length = 10.55 or $10\frac{9}{16}$"

Practice 34: Find the angle of the elbows and the cut length of the pipe for these offsets. Use a $\frac{3}{32}$" gap. Round the angle to the nearest half degree.

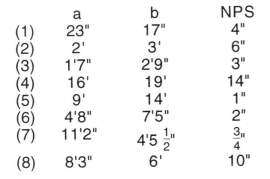

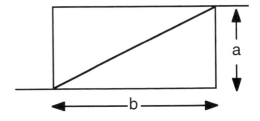

	a	b	NPS
(1)	23"	17"	4"
(2)	2'	3'	6"
(3)	1'7"	2'9"	3"
(4)	16'	19'	14"
(5)	9'	14'	1"
(6)	4'8"	7'5"	2"
(7)	11'2"	4'5 $\frac{1}{2}$"	$\frac{3}{4}$"
(8)	8'3"	6'	10"

There is a chart of factors on page 147 for the shortest simple offsets. The shortest simple offsets consist of a pair of equal odd angle elbows welded together. To find the angle of the elbows needed, divide the offset distance by the nominal size of the pipe you are running. Compare the answer to the factors in the chart and find the closest match. Use that angle as the angle for your elbows.

Combination Offsets

*A **combination offset** uses a 90° elbow* and another angle elbow, in order to accomplish an offset and a change of direction in one shot.*

90° - 45° Combination Offset

The most often used combination offset is the 90°-45°. It is calculated exactly the same way as a 45° simple offset, except for the take outs. Here are two views.

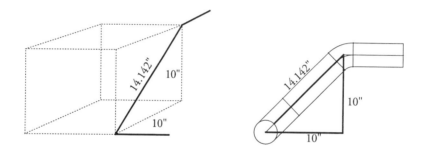

Look at the second drawing. The cut length is calculated by subtracting one 90° take out and one 45° take out, instead of two 45° take outs.

Practice 35: Find the cut length of the pipe for these 90°-45° offsets. Don't forget to subtract the $\frac{3}{32}$" gaps. Use field cut 45° elbows.

	Offset	NPS		Offset	NPS
(1)	22"	3"	(5)	34"	6"
(2)	2'3"	8"	(6)	7'4"	16"
(3)	5'7"	12"	(7)	6'	10"
(4)	10'	4"	(8)	12'3$\frac{3}{8}$"	1"

---

* A tee can be used instead of the 90° elbow if needed.

The Shortest 90°-45° Combination Offset

How much would a 90°-45° combination, with the elbows welded to each other, offset an 8" line of pipe? When a question like this is asked, think of what you already know from the information in the question. The take out for an 8" 90° B.W. elbow is 12", and the take out for an 8" 45° B.W. ell is 5". The sum of the take out of the two elbows is 17", and the elbows are set at a 45° angle.

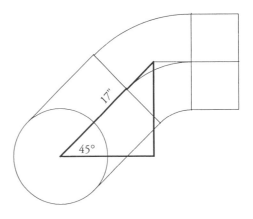

That is enough information to calculate the amount of offset.

Opposite = sin 45° x Hypotenuse

opposite = .7071 x 17"

Opposite = 12.02"

This is the smallest offset that a combination 90°-45° for 8" pipe can be. If you need a smaller offset, use a combination of a 90°- odd angle elbow.

90° - Odd Angle Combination Offset

Calculating a 90° - odd angle combination offset is the same as calculating a simple odd angle offset, except for the take outs. In this drawing, you need to figure the angle of rise to find the angle of the elbow at the top of the offset. The bottom elbow is a 90° elbow that will be turned up at an angle, equal to the angle of rise.

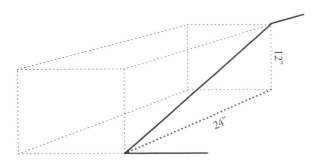

First, find the angle of rise using the two sides that are known.

$$\text{Tangent} = \frac{\text{opposite}}{\text{adjacent}} = \frac{12}{24} = .5 \boxed{\tan^{-1}} = 26.5651°$$

The angle of rise is 26.5°. The complementary angle is 63.5°.

The angle of rise is the angle of the elbow to be used in combination with the 90° elbow.

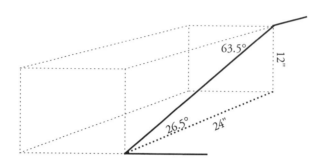

The hypotenuse of this triangle is 26.832".

Consider this an 8" line of pipe.

The take out for a $26\frac{1}{2}°$ elbow is 2.826".

Cut length = Run - T.O. for 90° ell - T.O. of $26\frac{1}{2}°$ ell - 2 gaps($\frac{3}{32}$)

Cut length = 28.832" - 12 - 2.826" -.1875

Cut length = 13.819" or $13\frac{13}{16}$"

What angle do you use to set the 90° elbow? It is almost impossible to get a correct reading by placing the angle finder on the top of an elbow, but it is easy to put your angle finder on the face of the elbow. The angle you measure this way is different than the angle of rise, but the angles are related as you will see. Take a look at the two right triangles in the drawing below.

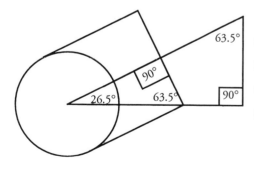

Since the center line coming out of the elbow is perpendicular to the face, a right triangle can be created. The face of the 90° elbow will be at a 63.5° angle, which is the complementary angle of the angle of rise.

When you are fitting an odd angle elbow, the face of the elbow will read as the complementary angle of the angle of the rise when the pipe is in the jack stands.

Practice 36: Find the angle of the elbows and the cut length of the pipe for these combination 90° - odd angle offsets. Use a $\frac{3}{32}$" gap, and round your angles to the nearest half degree.

	a	b	NPS
(1)	32"	13"	4"
(2)	11'	9'	6"
(3)	3'7"	2'9"	3"
(4)	26'	19'	14"
(5)	19'	12'	1"
(6)	8'8"	7'5"	2"
(7)	10'2"	4'5.75"	$\frac{3}{4}$"
(8)	8'4"	6'	10"

The Shortest Combination Offsets

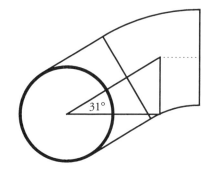

The shortest combination offsets are made by welding an odd angle elbow to a 90° elbow. The chart of factors below can be used to find the correct odd angle elbow to combine with a 90° elbow in order to achieve a combination that will give you the shortest offsets. If you divide the offset needed by the nominal pipe size and compare the answer with the factors in the chart, it will give you the angle of the elbow needed to combine with a 90° elbow for that offset(except in the case of the $\frac{1}{2}$" pipe).

This sample table has 5° divisions. There is a full table in the back of the book.

90° +	θ ell	Offset Factor
90° +	5°	0.1364
90° +	10°	0.2833
90° +	15°	0.439
90° +	20°	0.6035
90° +	25°	0.7745
90° +	30°	0.9510
90° +	35°	1.1316
90° +	40°	1.3151
90° +	45°	1.5000
90° +	50°	1.6849
90° +	55°	1.8684
90° +	60°	2.0490
90° +	65°	2.2255
90° +	70°	2.3965
90° +	75°	2.5607
90° +	80°	2.7167
90° +	85°	2.8636
90° +	90°	3.0000

For example: You are running an 8" line of pipe and need to turn 90° and rise $2\frac{1}{4}$". Divide $2\frac{1}{4}$" by 8

$$\frac{2.25}{8} = .28125$$

The closest factor is the factor with the 10° elbow. That means that a combination 90° elbow and a 10° elbow would give you the correct offset.

Practice 37: Find the combination of elbows that allows the offset shown with the listed pipe size.

	NPS	Offset		NPS	Offset
(1)	10"	9.5"	(5)	8"	21.75"
(2)	2"	2.25"	(6)	4"	1.75"
(3)	3"	6.125"	(7)	12"	18"
(4)	14"	23.5625"	(8)	6"	12.5"

Rolling Offsets

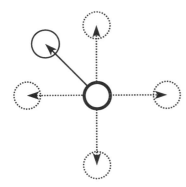

*The **rolling offset** is a simple offset that is rolled over to one side or the other.* It can be better understood if placed in an **offset box.** The rolling offset is a simple offset that runs diagonally through the box.

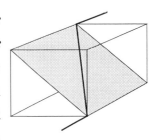

The Rolling Offset Box

You draw the box to keep the sides straight. Other than that, the box has no importance. You may be calculating an offset from a riser or a level pipe, but the offset can still be drawn in the standard box. The different offsets could look like these examples.

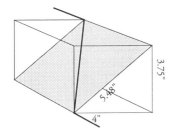

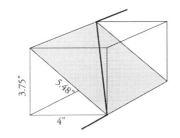

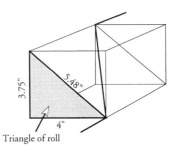

I draw my boxes like the one below, and name one of the triangles in it.

This triangle is the **Triangle of Roll**. It is usually the first triangle calculated in a rolling offset. The hypotenuse of the triangle of roll is a common side of the second triangle to be calculated in the 45° rolling offsets and the odd angle offsets.

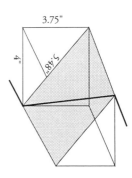

Triangle of roll

Notice that the lines entering and leaving the offset box are lined up with the lines of an edge of the box.

90° Rolling Offsets

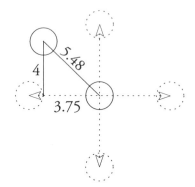

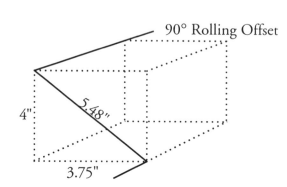

In a 90° rolling offset, the run of pipe crosses the face of the box diagonally. The length of the run is the hypotenuse of the triangle of roll. The pipe size of this 90° rolling offset is 1".

$$a^2 + b^2 = c^2$$
$$\sqrt{a^2 + b^2} = c$$
$$\sqrt{3.75^2 + 4^2} = c$$
$$\sqrt{14.0625 + 16} = c$$
$$\sqrt{30.0625} = c$$
$$5.4829 = c$$

Cut length = 5.48"- 2 take outs - 2 gaps ($\frac{3}{32}$")
Cut length = 5.48" - 2(1.5") - 2(.09375")
Cut length = 2.2925"

Practice 38: Find the cut length of these 90° rolling offsets. Use a $\frac{3}{32}$" gap

	NPS	a	b
(1)	6"	12"	17"
(2)	12"	6.5'	13"
(3)	2"	14'	11"
(4)	3"	43"	51"
(5)	8"	21"	22"
(6)	1"	7'	1.5'

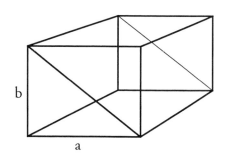

45° Rolling Offset

Using the same dimensions as the previous offset, here is a rolling offset with 45° elbows.

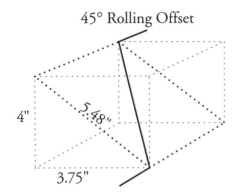

45° Rolling Offset

4"

5.48"

3.75"

A 45° rolling offset is a 45° simple offset that is rolled to one side.

Let's look at the simple offset without the roll.

Notice that the side measurement is the same as the hypotenuse of the triangle of roll. From your study of 45° right triangles, you know that the legs of a 45° right triangle must be equal; therefore, all four sides of the box are equal.

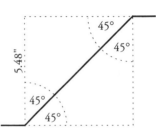

The run of pipe will be the length of the hypotenuse of the 45° right triangle.

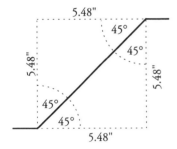

You know from an earlier section that a leg of a 45° right triangle x 1.4142 equals the hypotenuse.

5.48" x 1.4142 = 7.7498"

The run equals 7.7498".

To determine the cut length, consider this a 1" line of pipe, and use **factory made 45°** elbows this time!

To get the cut length, subtract the take outs for two 45° elbows and two gaps from the run. You can find the take out for a 45° elbow in the take out section.

Cut length = Run - 2 T.O. - 2 gaps

Cut length = 7.7498 - 2(.875) - 2(.09375)*

---

* When you find a number or term beside another term or number in parentheses, it means to multiply.   2(.875) = 2 x .875 = 1.75   2(.09375) = 2 x .09375 = .1875

Cut length = 7.7498 -1.75 -.1875

Cut length = 5.8123" or $5\frac{13}{16}$"

Let's take another look at what was needed. We started with two dimensions, basically 3.75" over and 4" up. They were used to find the hypotenuse length of the triangle of roll, 5.48". Since 45° elbows were to be used, the hypotenuse of the triangle of roll became a leg of a 45° right triangle. To find the length of the hypotenuse of the 45° right triangle, the leg of the 45° triangle (the hypotenuse of the triangle of roll) was multiplied times 1.4142, and the results was 7.75". The calculations for cut length were made according to the pipe size.

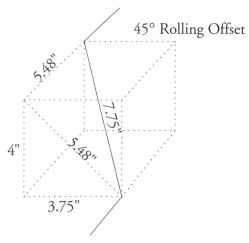

Practice 39: Find the cut length of these 45° rolling offsets using field cut elbows and a $\frac{1}{8}$" gap.

	a	b	NPS
(1)	27"	13"	3"
(2)	2.5'	4'	8"
(3)	1'9"	2'7"	4"
(4)	6'	11'	16"
(5)	8'	13'	6"
(6)	5'8"	7'2"	2"
(7)	11'7"	$4'3\frac{1}{2}$"	$\frac{3}{4}$"
(8)	7'4"	5'	10"

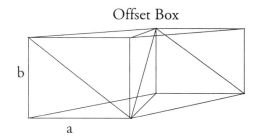

Offset Box

Odd Angle Rolling Offsets

Again, using the same dimensions as in the previous examples, let's look at an offset using two 30° elbows. The hypotenuse of the triangle of roll is the same. The length of the run is needed. There are two knowns in this case, an angle of 30° and the length of one side (5.48"). Which side of the triangle is this, the opposite or the adjacent?

The use of 30° elbows involves a 30° - 60° right triangle.

Is the 30° angle A or B?

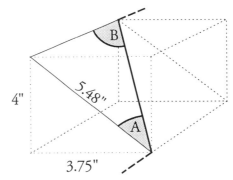

Below is a view of the simple offset of this rolling offset taken from the box. Look at the dimensions on the box and compare them to the rectangle below to orient yourself. Notice that the hypotenuse of the triangle of roll is again a side measurement of the next triangle to be calculated.

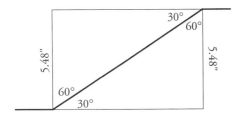

Look at the angles at the entry and exit points. The angle of the turn is behind the elbow, so the angle of turn (30°) in the offset box is angle B.

Your knowns are the opposite side (5.48") and the reference angle (30°). You need the length of the hypotenuse.

The chart for finding the sides shows that:

csc θ x opp = hypo (csc 30° = 2)

2 x 5.48" = 10.96".

H x sin θ = O	O x csc θ = H
H x cos θ = A	A x sec θ = H
A x tan θ = O	O x cot θ = A

Use a 1" pipe for this example and find the cut length of the pipe.

First, find the take out for a 30° elbow.

Take out = tan $\dfrac{\theta}{2}$ x radius

Take out = tan 15° x 1.5"   tan 15° = .2980

Take out = .2680 x 1.5"

Take out for a 30° B.W. elbow = .402"

Cut length = Run - 2 T.O. - 2 gaps ($\frac{3}{32}$" each)

Cut length = 10.96" - 2(.402") - 2(.09375)

Cut length = 10.96 - .804 - .1875

Cut length = 9.9685" or rounded off to 10"

Practice 40: Find the cut length of the pipe for these offsets using the angle of the elbows and dimensions given. Use a $\frac{1}{8}$ " gap

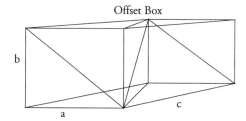

Offset Box

	a	b	c	angle of ell	NPS
(1)	14"	27"	52.679"	30°	3"
(2)	25"	43"	59.278"	40°	8"
(3)	2'	5'	177.5507"	20°	4"
(4)	4' 5"	6' 9"	55.8891"	60°	16"
(5)	8"	13"	12.8087"	50°	6"
(6)	4'3"	9'7"	179.6670"	35°	2"
(7)	12 11"	2' 5"	309.4892"	27°	$\frac{3}{4}$"
(8)	6'3"	4' 7"	286.2462"	18°	10"

Angle of Turn and Rise of a Rolling Offset

You have seen the terms *angle of turn and angle of rise* used with the simple offsets. They are also used with the rolling offset, but they are viewed differently. You will need them in the next section, so let's show how they work using the last example.

The Angle of Turn

The angle of turn for a rolling offset is still found behind the elbow, but it is found on the bottom of the box. The dimensions of the triangle on the bottom of the box are used to find the angle of turn.

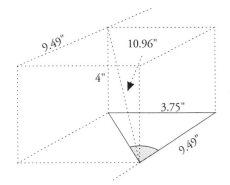

$$\text{Tangent} = \frac{3.75}{9.49} = .3952 \boxed{\tan^{-1}} = 21.56°$$

The Angle of Rise

The angle of rise is the angle the diagonal rises from the floor of the offset box.

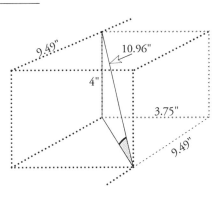

$$\text{Sine} = \frac{4}{10.96} = .3650 \boxed{\sin^{-1}} = 21.41°$$

The angles of turn and rise are used in determining the angle of the elbows to be used in the offsets.

If the cosine of the angle of rise is multiplied times the cosine of the angle of turn, the answer will be the cosine of the angle of the elbow that you need.

Let's see how it works.

Angle of rise 21.41° $\boxed{\cos}$ =.9310

Angle of turn 21.56° $\boxed{\cos}$ =.9300

.9310 x .9300 =.86583 $\boxed{\cos^{-1}}$ = 30.02°

The angle of the elbows used in the offset above is 30°.

Use Practice 40 in the section on odd angle rolling offsets to sharpen your skills on finding the angles of the elbows. You should come out with the elbows listed in the practice. You might be off a slight amount due to the conversions, but it should round off to the correct elbow angle.

Combination Rolling Offsets

The difference between the **combination rolling offsets** and the other rolling offsets is that the combination rolling offsets change direction as well as roll. If you go back and look at the drawings of the other rolling offsets, you will see that the entry and exit lines are all *parallel**. In the examples below you will see that the exit lines leave at a different angle from the entry lines. *The combination rolling offset always uses two different degree elbows, but unlike the other combination offsets, it never uses a 90° elbow.*

As you can see, the exit line is not parallel to the entry line, but is at a 90° turn from the entry line after the roll.

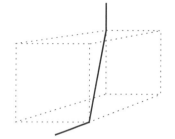

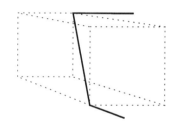

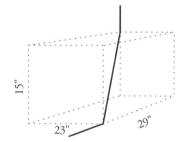

You need to know three dimensions—length, width, and depth—in order to work a combination rolling offset. You also need to know the direction in which the exiting pipe leaves the offset. You will have to calculate both elbows, but the method of calculation depends on the direction in which the exiting line leaves the offset box.

First, find the angle of rise and the angle of turn. You will also need to find the length of the run.

---

* Parallel lines are lines in the same *plane* that never meet no matter how far you extend them.

To find the angle of turn, you can use the tangent function.

$\text{Tan } \theta = \dfrac{\text{opp}}{\text{adj}} = \dfrac{23}{29} = .7931 \boxed{\tan^{-1}} = 38.42°$

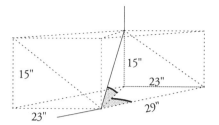

At this point, the triangle containing the angle of rise has only 1 known side; however, it does have a **common side** with the triangle containing the angle of turn. If you find the hypotenuse of that triangle, you will also have the length of the adjacent side of the triangle containing the angle of rise.

$$a^2 + b^2 = c^2$$
$$\sqrt{a^2 + b^2} = c$$
$$\sqrt{23^2 + 29^2} = c$$
$$\sqrt{529 + 841} = c$$
$$\sqrt{1370} = c$$
$$37.01 = c$$

Now you have the length of the hypotenuse of the first triangle and the length of the adjacent leg of the next triangle.

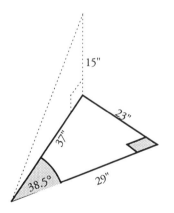

Look at what has been done so far on these triangles which have been taken from the box. You can see the common side of the two triangles (37").

You can now use the tangent function to find the angle of rise.

$\text{Tan } \theta = \dfrac{\text{opposite}}{\text{adjacent}} = \dfrac{15}{37} = .40541 \boxed{\tan^{-1}} = 22.07°$

Next, find the length of the run or the hypotenuse of the second triangle.

$$\sqrt{a^2 + b^2} = c$$
$$\sqrt{37^2 + 15^2} = c$$
$$\sqrt{1369 + 225} = c$$
$$\sqrt{1594} = c$$
$$39.93 = c$$

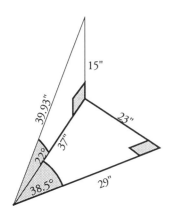

You are now at the point where you can find the angle of the entry elbow. *You find the angle of the elbow by multiplying the cosine of the angle of turn times the cosine of the angle of rise, which will give you the cosine of the angle of the elbow.*

Cos 22° x cos 38.5° = cos θ of angle of elbow

.9272 x .7826 = cos θ of angle of elbow

.7256 = cos θ of angle of elbow

.7256 $\boxed{\cos^{-1}}$ = 43.48°

The entry elbow has an angle of $43\frac{1}{2}°$.

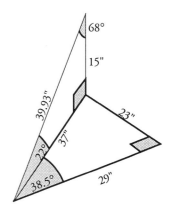

The upper angle is a different story, because it rises straight up out of the box without a turn. The upper elbow is the complement angle of the angle of rise, so the upper elbow has an angle of 68°.

Using the same dimensions, let's turn the exiting line in a different direction.

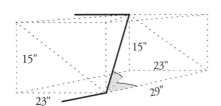

All of the triangles are the same, and the bottom elbow has the same angle. The only thing that has changed is the angle of the exiting elbow. It must be found in the same way that the entry elbow was. The angle of rise is the same, but the angle of turn will be the complement of the angle of turn for the entry elbow.

Take a look.

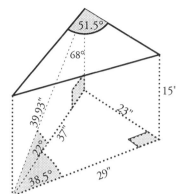

The angle of turn for the exiting elbow is $51\frac{1}{2}°$.
The angle of rise is still 22°.
The cosine of $51\frac{1}{2}°$ is .6225.
The cosine of 22° is .9272.

.6225 x .9272 = .5772 $\boxed{\cos^{-1}}$ =54.75° The exiting elbow will be a 55° ell.

The process of determining the cut length for combination rolling offsets differs from that for regular rolling offsets only by having two different take outs included in each problem instead of one take out that is doubled.

Practice 41: In each of the following problems, find the cut length and the angle for each elbow. Use a $\frac{3}{32}$" gap.

Remember to round off the angles of the elbows to the nearest $\frac{1}{2}$ before calculating cut length.

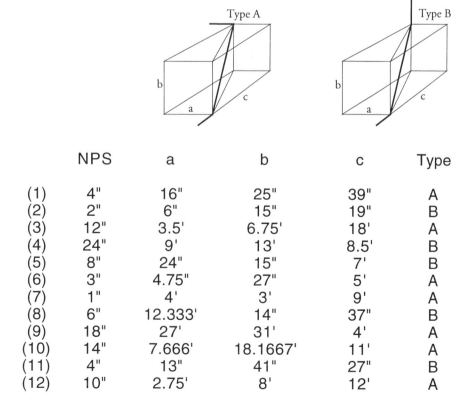

	NPS	a	b	c	Type
(1)	4"	16"	25"	39"	A
(2)	2"	6"	15"	19"	B
(3)	12"	3.5'	6.75'	18'	A
(4)	24"	9'	13'	8.5'	B
(5)	8"	24"	15"	7'	B
(6)	3"	4.75"	27"	5'	A
(7)	1"	4'	3'	9'	A
(8)	6"	12.333'	14"	37"	B
(9)	18"	27'	31'	4'	A
(10)	14"	7.666'	18.1667'	11'	A
(11)	4"	13"	41"	27"	B
(12)	10"	2.75'	8'	12'	A

Notes

XI

Answers to the Practices

Practice 1   (page 14)

| (1) | $\frac{3}{8}$ | (2) | $\frac{1}{4}$ | (3) | $\frac{1}{4}$ | (4) | $\frac{3}{4}$ | (5) | $\frac{3}{4}$ | (6) | $\frac{1}{2}$ | (7) | $\frac{7}{8}$ | (8) | $\frac{3}{4}$ |

Practice 2   (page 16)

(1)	.375"	(2)	.6875"	(3)	.4375"	(4)	.5"
(5)	.75"	(6)	.59375"	(7)	.875"	(8)	.1875"
(9)	.125"	(10)	.25"	(11)	.9375"	(12)	.0625"

Practice 3   (page 17)

(1)	17.75"	(2)	43.9375"	(3)	23.6875"	(4)	3.1875"
(5)	7.125"`	(6)	87.875"	(7)	2.03125"	(8)	4.625"

Practice 4   (page 18)

(1)	$5\frac{7}{12}' = 5.5833'$	(2)	$4\frac{1}{12}' = 4.0833'$	(3)	$92\frac{5}{12}' = 92.4167'$
(4)	$19\frac{9}{12}' = 19.75'$	(5)	$46\frac{10}{12}' = 46.8333'$	(6)	$137\frac{8}{12}' = 137.6667'$

Practice 5   (page 19)

(1)	6.5'	(2)	1.8333'	(3)	3.4792'	(4)	16.6719'
(5)	23.3854'	(6)	123.0313'	(7)	42.7969'	(8)	1.9844'
(9)	96.2708'	(10)	37.4948'				

Practice 6   (page 21)

(1)	$\frac{7}{16}"$	(2)	$\frac{5}{8}"$	(3)	$\frac{5}{16}"$	(4)	$\frac{9}{16}"$
(5)	$\frac{15}{16}"$	(6)	$\frac{3}{16}"$	(7)	$\frac{13}{16}"$	(8)	$\frac{5}{16}"$

Practice 7:   (page 22)

(1)	$12\frac{3}{8}$"	(2)	$4\frac{5}{8}$"	(3)	$99\frac{15}{16}$"	(4)	$56\frac{3}{4}$"
(5)	$74\frac{3}{16}$ "	(6)	$19\frac{9}{16}$"	(7)	$37\frac{7}{16}$"	(8)	$23\frac{1}{16}$"

Practice 8: (page 23)

(1)	24' 9"	(2)	$1' 10\frac{13}{16}$"	(3)	3' 4"	(4)	$170' 10\frac{3}{8}$"
(5)	$53' 1\frac{1}{2}$ "	(6)	$42' 8\frac{15}{16}$ "	(7)	$33' 10\frac{3}{4}$ "	(8)	$75' 1\frac{9}{16}$ "
(9)	$84' 5\frac{1}{4}$ "	(10)	$4' 6\frac{13}{16}$ "	(11)	$8" 3\frac{9}{16}$ "	(12)	$12' 11\frac{15}{16}$ "

Practice 9   (page 32)

(1)	100	(2)	529	(3)	52900	(4)	306.25
(5)	30.9414	(6)	87.8906	(7)	121	(8)	625

Practice 10 (page 34)

(1)	1.4142	(2)	8.3066	(3)	3.8079	(4)	4.2131
(5)	16	(6)	32.2025	(7)	9.7243	(8)	5.9055

Practice 11 (page 38)

(1)	15	(2)	25	(3)	39.1152"	(4)	8.1432"
(5)	11.3724"	(6)	8.4364'	(7)	14.5765'	(8)	32.0355"
(9)	128.5518'	(10)	30	(11)	29.9125"	(12)	51

Practice 12: (page 40)

(1)	9	(2)	14.6225	(3)	17.6635	(4)	7.8727'
(5)	62.5780'	(6)	19.4936"	(7)	10.1698	(8)	43.0414
(9)	15.4919	(10)	19.1242	(11)	8.6603'	(12)	1.7199'

Practice 13: (page 44)

Angle A		Angle B	
Sin θ = .4223	csc θ = 2.3681	Sin θ = .9067	csc θ = 1.0286
cos θ = .9067	sec θ = 1.0286	cos θ = .4223	sec θ = 2.3681
tan θ = .4657	cot θ = 2.1472	tan θ = 2.1472	cot θ = .4657

Angle C		Angle D	
sin θ = .7071	csc θ = 1.4142	sin θ = .7071	csc θ = 1.4142
cos θ = .7071	sec θ = 1.4142	cos θ = .7071	sec θ = 1.4142
tan θ = 1	cot θ = 1	tan θ = 1	cot θ = 1

Practice 14: (page 47)

(1) 25°	(2) 45°	(3) 70°	(4) 90°
(5) 30°	(6) 61°	(7) 20°	(8) 64°

Practice 15: (page 49)

(1)	$40\frac{1}{2}$° & $49\frac{1}{2}$°	(2)	55° & 35°	(3)	34° & 56°
(4)	26° & 64°	(5)	56.5° & 33.5°		

Practice 16: (page 52)

(1)	28°/62°	(2)	53°/37°	(3)	28°/62°	(4)	56.5°/33.5°

Practice 17: (page 53)

(1)	tan 45° = 1	(2)	sin 33° = .5446	(3)	cos 25° = .9063
(4)	sin 60° = .8660	(5)	cos 54° = .5878	(6)	tan 75° = 3.7321
(7)	sin 89° = .9998	(8)	cos 60° = .5		

Practice 18: (page 55)

(1)	csc 39° =1.589	(2)	sec 45° = 1.4142	(3)	cot 30° = 1.7321
(4)	sec 15° = 1.0353	(5)	cot 60° = .5774	(6)	csc 23° = 2.5593
(7)	sec 55° = 1.7434	(8)	cot 1° = 57.29		

Practice 19: (page 56)

(1)	85°	(2)	89°	(3)	37°	(4)	30°
(5)	25°	(6)	48°	(7)	20°	(8)	80°

Practice 20: (page 59)

(1)	Hypo = 3.3101" Adj = 1.3989"	(5)	Hypo = 6.7340' Opp = 3.0572'
(2)	Opp = 2.1213' Adj = 2.1213'	(6)	Hypo = 2.5380' Adj = 1.5626'
(3)	hypo = 11.1697" opp = 1.9396"	(7)	Adj = 3.3646" opp = 12.5570"
(4)	hypo = 57.29869" opp = 57.28996"	(8)	Adj = 5.1289' opp = 97.8657'

Practice 21: (page 66)

(1)	$6'3\frac{3}{8}$ "	(2)	11"	(3)	31"	(4)	$21' 2\frac{1}{2}$ "
(5)	$15\frac{11}{16}$ "	(6)	$24\frac{3}{8}$ "	(7)	31' 5"	(8)	$78' 6\frac{1}{2}$ "

Practice 22: (page 67)

(1)	$4\frac{1}{8}$ "	(2)	11 "	(3)	$27\frac{1}{8}$ "	(4)	$40\frac{1}{16}$ "	(5)	44"	(6)	$75\frac{3}{8}$ "

Practice 23: (page 68)

(1)	$6\frac{5}{16}$"	(2)	$157'\frac{15}{16}$"	(3)	$81\frac{11}{16}$"	(4)	647' 2"
(5)	$13\frac{3}{8}$"	(6)	$6'3\frac{3}{8}$"	(7)	$10'\frac{15}{16}$"	(8)	$376'11\frac{7}{8}$"

Practice 24: (page 73)

(1) 10° = .17453	(2) 48° = .8378	(3) 57° = .9948	(4) 65° = 1.1345
(5) 72° = 1.2566	(6) 39° = .6807	(7) 22° = .3840	(8) 84° = 1.4661

Practice 25: (page 74)

(1)	$5\frac{5}{16}$"	(2)	5"	(3)	$3\frac{9}{16}$"	(4)	$4\frac{3}{16}$"
(5)	$8\frac{1}{2}$"	(6)	$3'0\frac{7}{8}$"	(7)	$4'7\frac{5}{16}$"	(8)	$2\frac{3}{4}$"
(9)	$105'6\frac{11}{16}$"	(10)	$3\frac{15}{16}$"	(11)	$4\frac{3}{16}$"	(12)	$47\frac{1}{8}$"

Practice 26: (page 79)

(1)	$2\frac{7}{8}$"	(2)	$3\frac{11}{16}$"	(3)	$4\frac{5}{32}$"	(4)	$3\frac{29}{32}$"

Practice 27: (page 82)

	Adjacent	Opposite		Adjacent	Opposite		Adjacent	Opposite		Adjacent	Opposite
(1)	.9808	.1951	(2)	.9239	.3827	(3)	.8315	.5556	(4)	.7071	.7071
(5)	.5556	.8315	(6)	.3827	.9239	(7)	.1951	.9808			

There are a few secrets in these answers to Practice 27.
One is that the lengths of the opposite and adjacent sides are the same as the sine and cosine functions for those respective angles. Another is that there are only 4 different triangles in these seven. Three of these seven are duplicates. The only one that does not have a duplicate is the 45° right triangle. Look at the answers above and see that three angles are complementary to three others. If all the hypotenuses are the same, and the angles are the same then the triangles have to be the same size.

Extra Question from circles: The volume of a 6" B.W. elbow is 408.426 cubic inches, or 1.768 gallons. Use the center arc length as the length in the volume formula.

Practice 28: (page 91)

	inside	outside		inside	outside		inside	outside		inside	outside
(1)	$8\frac{1}{16}$"	$17\frac{1}{16}$"	(2)	$2\frac{15}{16}$"	$6\frac{1}{2}$"	(3)	$2\frac{1}{8}$"	$4\frac{7}{8}$"	(4)	$3\frac{11}{16}$"	$7\frac{13}{16}$"
(5)	$6\frac{1}{16}$"	$12\frac{3}{4}$"	(6)	$17\frac{1}{4}$"	$34\frac{9}{16}$"	(7)	$1\frac{1}{2}$"	$3\frac{1}{4}$"	(8)	$21\frac{1}{8}$"	$42\frac{1}{4}$"

Practice 29: (page 101)

(1)	$4\frac{1}{2}$"	(2)	$8\frac{3}{8}$"	(3)	$2\frac{11}{16}$"	(4)	$7\frac{1}{4}$"	(5)	$\frac{1}{2}$"
(6)	$1\frac{3}{8}$"	(7)	$8\frac{11}{16}$"	(8)	$2\frac{11}{16}$"	(9)	$3\frac{1}{8}$"	(10)	10"

Practice 30: (page 106)

(1)	$1'2\frac{5}{16}$"	$25°/65°$	(2)	$4'9\frac{7}{16}$"	$55°/35°$	(3)	$1'11\frac{5}{8}$"	$36\frac{1}{2}°/53\frac{1}{2}°$
(4)	$3'8\frac{1}{8}$"	$33°/57°$	(5)	$5\frac{1}{16}$"	$30\frac{1}{2}°/59\frac{1}{2}°$	(6)	$10\frac{5}{16}$"	$61°/29°$
(7)	$4'0\frac{7}{16}$"	$68°/22°$	(8)	$4'2\frac{15}{16}$"	$45°/45°$	(9)	$5'0\frac{3}{4}$"	$70°/20°$
(10)	$7'10\frac{7}{8}$"	$32\frac{1}{2}°/57\frac{1}{2}°$	(11)	$1'6\frac{9}{16}$"	$47\frac{1}{2}°/42\frac{1}{2}°$	(12)	$2'8\frac{1}{2}$"	$45°/45°$

Practice 31: (page 108)

(1)	$5\frac{3}{4}$"	(2)	$6'3\frac{3}{4}$"	(3)	$6\frac{1}{4}$"	(4)	$4\frac{3}{4}$"
(5)	$7\frac{3}{4}$"	(6)	$7'3\frac{3}{4}$"	(7)	$17'10\frac{3}{4}$"	(8)	$19'4\frac{1}{4}$"

Practice 32: (page 109)

(1)	$4'1\frac{5}{8}$"	(2)	$1'3\frac{7}{8}$"	(3)	$5'11\frac{1}{2}$"	(4)	$5'6\frac{7}{16}$"	(5)	$9'10\frac{9}{16}$"	(6)	$15'10\frac{1}{4}$"
(7)	$1'11\frac{1}{16}$"	(8)	$4'2\frac{5}{16}$"	(9)	$6'2\frac{1}{16}$"	(10)	$5\frac{5}{8}$"	(11)	$27'7\frac{5}{16}$"	(12)	$25'1$"

Practice 33: (page 110)

(1)	$3' 3\frac{1}{2}"$	(2)	$1' 5\frac{3}{4}"$	(3)	$6' 8\frac{3}{8}"$
(4)	$4' 1\frac{1}{4}"$	(5)	$8'4\frac{15}{16}"$	(6)	$11' 5\frac{7}{8}"$
(7)	$3' 5\frac{15}{16}"$	(8)	$4' 0\frac{7}{16}"$	(9)	$5' 1"$
(10)	$5\frac{5}{8}"$	(11)	$23' 0\frac{7}{8}"$	(12)	$19' 6\frac{3}{16}"$

Practice 34: (page 113)

(1)	$1' 10\frac{3}{8}"$	$53\frac{1}{2}°$	(2)	$3' 1\frac{11}{16}"$	$33\frac{1}{2}°$	(3)	$2' 11\frac{1}{2}"$	$30°$	(4)	$23' 6\frac{5}{8}"$	$40°$
(5)	$16' 6\frac{11}{16}"$	$32\frac{1}{2}°$	(6)	$8' 7\frac{1}{4}"$	$32°$	(7)	$11' 10\frac{9}{16}"$	$68°$	(8)	$8' 10\frac{15}{16}"$	$54°$

Practice 35: (page 114)

(1)	$2' 0\frac{9}{16}"$	(2)	$1' 9"$	(3)	$5' 9\frac{1}{8}"$	(4)	$13' 5\frac{1}{16}"$
(5)	$2' 11\frac{3}{16}"$	(6)	$7' 6\frac{5}{16}"$	(7)	$6' 8\frac{7}{16}"$	(8)	$17' 2\frac{1}{8}"$

Practice 36: (page 117)

(1)	$68°$	$2' 0\frac{5}{16}"$	(2)	$50\frac{1}{2}°$	$13' 1\frac{1}{8}"$	(3)	$52.5°$	$3' 11\frac{5}{16}"$	(4)	$54°$	$29' 6\frac{9}{16}"$
(5)	$57.5°$	$22' 3\frac{1}{8}"$	(6)	$49\frac{1}{2}°$	$11' 0\frac{5}{16}"$	(7)	$66°$	$10' 11\frac{1}{4}"$	(8)	$54°$	$8' 4\frac{3}{8}"$

Practice 37: (page 118)

(1)	$30°$	(2)	$35°$	(3)	$60°$	(4)	$50°$
(5)	$80°$	(6)	$15°$	(7)	$45°$	(8)	$61°$

Practice 38: (page 120)

(1)	$2\frac{5}{8}$"	(2)	3' $6\frac{7}{8}$"	(3)	13' $6\frac{3}{16}$"
(4)	4' $9\frac{1}{2}$"	(5)	$6\frac{1}{4}$"	(6)	6' $10\frac{3}{4}$"

Practice 39: (page 122)

(1)	3' $2\frac{3}{8}$"	(2)	5' $9\frac{7}{8}$"	(3)	3' $11\frac{3}{4}$"	(4)	16' $0\frac{1}{2}$"
(5)	20' $11\frac{5}{16}$"	(6)	12' $8\frac{5}{16}$"	(7)	17' $4\frac{7}{16}$"	(8)	11' $5\frac{15}{16}$"

Practice 40: (page 124)

(1)	4' $10\frac{3}{16}$"	(2)	5' $8\frac{3}{8}$"	(3)	15' $6\frac{9}{16}$"	(4)	6' $11\frac{13}{16}$"
(5)	$11\frac{5}{16}$"	(6)	18' $1\frac{3}{16}$"	(7)	28' $10\frac{9}{16}$"	(8)	24' 8"

Practice 41: (page 129)

	Cut Length	Entry ell	Exit ell		Cut length	Entry ell	Exit ell
(1)	3' $6\frac{1}{2}$"	37.5°	71°	**(7)**	10' 2"	29°	67°
(2)	1' $10\frac{1}{8}$"	40.5°	53°	**(8)**	11' $5\frac{3}{4}$"	76°	85°
(3)	17' $11\frac{11}{16}$"	23°	79.5°	**(9)**	38' $2\frac{5}{8}$"	84.5°	49°
(4)	14' $11\frac{7}{16}$"	61.5°	43.5°	**(10)**	20' $3\frac{11}{16}$"	61°	70°
(5)	6' $4\frac{5}{16}$"	18.5°	80.5°	**(11)**	3' $9\frac{5}{16}$"	58°	36°
(6)	5' $0\frac{5}{8}$"	24.5°	86°	**(12)**	13' $2\frac{7}{8}$"	35°	79°

Answers for the Self Help Test
Page 24

(1) $11\frac{5}{8}"$

(2) $5'11\frac{9}{16}"$

(3) $42.25'$

(4) $14.8125"$

(5) $16'8\frac{15}{16}"$

(6) $28.4323'$

(7) $10\frac{1}{8}"$

(8) $33.75"$

(9) $3'4\frac{1}{8}"$

(10) $148.9688'$

(11) $93'9\frac{1}{16}"$

(12) $23.875"$

(13) $72\frac{9}{16}"$

(14) $47.5052'$

(15) $89'8\frac{1}{8}"$

(16) $56\frac{7}{16}"$

(17) $1'0\frac{1}{16}"$

(18) $4.375"$

(19) $24\frac{15}{16}"$

(20) $47.4531'$

XII

Glossary

Term	Definition
Acute angle	Any angle between 0° and 90°.
Adjacent side	The side of a right triangle that is next to the reference angle.
Angle	1. The shape made by two straight lines meeting in a point. 2. The space between those lines. 3. The amount of space measured in degrees.
Angle finder	A tool used to find the angle of a line or surface.
Angle of drop	The angle that a line of pipe is directed toward the ground.
Angle of rise	The angle that a line of pipe is directed up.
Angle of turn	The angle that a line of pipe is directed to the side.
Arc function (also called inverse function)	Used to find the degrees on an angle when the ratio of the sides of the angle is known. The six arc functions are: **arcsine, arccosine, arctangent, arccosecant, arcsecant, and arccotangent.**
Arc	A curved line.
Arc length	The length of a curved line.
Area of a circle	The surface measurement of a circle, expressed in square units.
Back of the elbow	The outside arc of the elbow.
Bolt hole circle	A circle drawn through the center of the bolt holes of a flange.
Butt weld	A type of fitting in which the beveled end of the fitting is welded to the beveled butt of a pipe.
Center to end	The measurement from the center of an elbow to the face of the elbow.
Central angle	An angle in which the vertex is located at the center of a circle.
Chord	A straight line from one point on a circle to another point on the circle. The diameter is the longest chord of a circle.
Circle	A curved line in a plane that encloses a space, with every point of the curved line exactly the same distance from the center point.
Circumference	The distance around a circle or a pipe.
Combination offset	An offset of pipe that uses a 90° elbow and another degree elbow. It offsets and changes direction at the same time.
Combination rolling offset	An offset that accomplishes a roll and a turn. It is made by using two different angled elbows, with neither elbow being a 90° ell.

Common side of two triangles	A line that is shared by two triangles. It can be the hypotenuse of one triangle and the leg of the other.
Compass	A tool used to draw circles and curved lines.
Complementary angles	Two angles whose sum equals 90°. The two lesser angles of a right triangle are always complementary.
Cosecant	One of the functions of an angle. It is found by dividing the hypotenuse of a right triangle by the opposite side of a reference angle.
Cosine	One of the functions of an angle. It is found by dividing the adjacent side of a reference angle by the hypotenuse.
Cotangent	One of the functions of an angle. It is found by dividing the adjacent side of a reference angle by the opposite side of a reference angle.
Cut length	The length of a pipe needed between the elbows of an offset.
Decimal	Shortened version of decimal fraction.
Decimal fraction	A fraction in which the denominator is 10 or a power of 10 (such as 100, 1000, 10,000,etc). The denominator is seldom used; instead, a dot called a decimal point is placed in front of the numerator. The denominator can be inferred by the number of decimal places behind the decimal point. If there is one number to the right of the decimal point, the denominator is 10. If there two numbers to the right of the decimal point, the denominator is 100. There will be the same number of zeroes in the denominator as there are decimal places.
Degrees	The units of measure for angles.
Denominator	The bottom number in a fraction. It is equal to the portions that the whole is divided into. In the fraction $\frac{9}{16}$ ", the whole inch has been divided into 16 parts and the measurement is equal to 9 of them.
Diameter	A straight line from one side of a circle to the other side that passes through the center.
Division bar	The line between the numerator and the denominator of a fraction.
Entry point	The point at the corner of an offset box where the line of pipe enters the box. If the offset is made with elbows, the entry point will be the center of the elbow.
Exit point	The point at the corner of an offset box where the line of pipe leaves the box. If the offset is made with elbows, The exit point will be the center of the elbow.
Face	Can be either the face of a pipe or the face of an elbow. It is the end of the pipe or the end of the fitting.
Face to face	The measurement from the face of one elbow of an offset to the face of the other elbow.
Factory made elbows	90° or 45° elbows that are made to the standards set by the American Society of Mechanical Engineers (ASME). They are singled out in this book because of the difference in the take outs between the field cut and factory made 45° B.W. elbows.
Field cut elbows	Elbows cut from 90° elbows on the job site.
Flanges	Fittings used to bolt pipes to fittings or fittings to other fittings.
Formula	An equation that follows a rule. Formulas contains variables that are replaced by numbers when they are known to find unknown quantities. An example of a formula is the one from the Pythagorean Theorem, $a^2 + b^2 = c^2$.
Fraction	A way of expressing a portion of a whole. The denominator represents the portions the whole has been divided into, and the numerator expresses the number of the portions measured. The fraction $\frac{1}{4}$ could be stated as 1 of 4 parts of the whole.

Function of an angle	The name of a particular ratio of the sides of a right triangle. The name of the function will depend on which side is divided into which side. There are six possible ways to divide the sides of a right triangle into each other. There are six functions for each angle. The names are: sine, cosine, tangent, cosecant, secant, and cotangent.
Functions table	A table of the functions of angles. This table allows the viewer to find either the function of an angle or the degree of an angle.
Hypotenuse	The longest side of a right triangle. It is always located directly across from the right angle.
Inside arc of an elbow	The smallest arc of an elbow.(also know as the throat)
Inside Diameter (I.D.)	The diameter of the inside wall of a pipe.
Inverse	Means turned upside down. There are many other ways it can be used, but let's not cloud the issue.
Isosceles right triangle	A 45° right triangle. There is only one right triangle that has two equal legs – the 45° right triangle. There is only one right triangle that has equal lesser angles – the 45° right triangle.
Isosceles triangle	A triangle that has two equal legs and two equal angles.
Jack stands	Devices used to hold the pipe at a working level. They are usually adjustable and are used in pairs.
Knowns	Variables in formulas that can be replaced by the actual measurements.
Lateral	A term, in pipe fitting, for a pipe entering a header at an angle other than perpendicular.
Legs of a right triangle	The sides of a right triangle other than the hypotenuse.
Make up	The distance a pipe is inserted into a fitting
Minutes	A division of a degree. There are 60 minutes in each degree.
Miter	To cut a pipe at an angle.
Mixed number	A whole number and a fraction.
Needs	The unknowns of a formula, or the measurements needed to do a fit.
Nominal pipe size (NPS)	The size by which a pipe is called. For example, the NPS of a schedule 40 pipe that has an O.D. of 12.75" and an I.D. of 11.9375 is 12".
Numerator	The top number of a fraction. It indicates the number of portions of a whole.
Obtuse angle	An angle between 90° and 180°.
Odd angle elbow	Any angle elbow besides a 90° elbow or a 45° elbow.
Offset	Usually a combination of two elbows and a cut length of pipe that moves a line of pipe to a new position.
Opposite side	The side of a right triangle that is opposite the reference angle.
Outside arc	The longest arc of an elbow.(also known as the back of an elbow)
Outside diameter (O.D.)	The measurement from one outside edge of a pipe to the other outside edge, going through the center point.
Parallel lines	Lines that are in the same plane and will always stay the same distance apart.
Perpendicular	A 90° angle between two lines. It is also called a **right angle**.
Pi (π)	The ratio of the circumference of a circle to its diameter. It is the whole number 3 and a decimal fraction that has endless decimal places. We usually round it off to four decimal places for our work.

Plumb	Exactly vertical. A line that perpendicular to ground level is plumb.
Radians	A measurement of arc length based on the length of the radius. One **radian** is an arc equal in length to the radius. The central angle that cuts off this arc is $\dfrac{180°}{\pi}$, or 57.295+°.
Radius	A straight line from the center of a circle to a point on the circle.
Radius of the elbow	The distance from the vertex of an elbow to the center line of the elbow. For a long radius elbow, it is $1\frac{1}{2}$ times the nominal pipe size with the exception of the $\frac{1}{2}$" elbow.
Ratio	A comparison of one value to another value by division.
Ratios of the sides	The division of one side of a right triangle by another side. The ratios of the sides are directly related to the number of degrees in the reference angle.
Reference angle	The angle being referred to in any given problem.
Right triangle	A triangle with one angle equal to 90° and two angles whose sum is 90°.
Rolling offset	A simple offset that is rolled to one side.
Rolling Offset Boxes	Boxes that are drawn by the fitter to keep the sides and length of sides straight when calculating a rolling offset.
Rounding off a number	Appropriate accuracy. There is a point at which the accuracy of your work is restricted by the tools you are working with. Your calculations do not need to be any finer than that, so you round the number off at that point. An example is rounding off to .5°. We do not have the tools to be more accurate than that; therefore, our calculations need not be any more accurate than that.
Run	The path that a pipe takes to get from one center line to another center line.
Screwed elbow	An elbow that is joined to the pipe by threads.
Secant	One of the functions of an angle. It is found by dividing the hypotenuse of a right triangle by the adjacent side of a reference angle.
Seconds	A division of the minutes of a degree. There are 60 seconds in each minute of a degree.
Simple offset	Moving of the center line of a line of pipe from one location to another using a distance which can be measured.
Sine	One of the functions of an angle. It is found by dividing the opposite side of a reference angle by the hypotenuse of the right triangle.
Socket weld elbows	Elbows that are connected to the pipe, by inserting the pipe into the elbow, then welding them together.
Square root	A factor of a number, when squared gives that number. A square root asks what number times itself equal the number under the square root symbol.
Square	The product of a number multiplied times itself.
Stab in	A slang expression for a branch line entering a header without a pre made fitting.
Straight angle	An angle of 180° (in other words a straight line).
Take out	The distance that a fitting extends the center line of a run of pipe past the end of the pipe.
Take out formula	A formula to find the take out of an elbow. The take out formula is: **Take out = Tan $\dfrac{\theta}{2}$ x radius of the elbow.**
Tangent	One of the functions of an angle. It is found by dividing the opposite side of a reference angle by the adjacent side of a reference angle.

Theta - θ	A letter of the Greek alphabet that is used as a variable for unknown angles or when referring to any angle.
Throat of an elbow	The smallest arc of an elbow.
Triangle of roll	a triangle in a rolling offset box that contains the angle of roll. It is usually the first triangle calculated.
Triangle	an enclosed geometric form with three sides. The angles of a triangle always equal 180°
Unit circle	a circle with a radius of 1 unit.
Variables	letters of the alphabet used as symbols to represent numbers that change.
Vertex	the point at which two straight lines come together to form the angle.
Welder's gap	the space between a fitting and the pipe that is filled by welding.
Wraparound	a tool for marking straight lines on pipe.
Zero degree angle	An angle with no space between the two lines. In other words, the two lines are occupying the same space. The difference between a straight angle and a zero degree angle is that the vertex is in the middle of the line for a straight angle but at the end of the line for a zero degree angle.

Notes

XIII

Tables and Charts

The Dimensions for 90° Long Radius Butt Weld Elbows

Nominal Pipe Size	Pipe O D	Center line Radius	Inside Radius	Outside Radius
0.75"	1.05"	1.50"	0.975"	2.025"
1"	1.315"	1.5"	0.8425"	2.1575"
1.25"	1.66"	1.875"	1.045"	2.705"
1.5"	1.9"	2.25"	1.3"	3.2"
2"	2.375"	3"	1.812"	4.1875"
2.5"	2.875"	3.75"	2.312"	5.1875"
3"	3.5"	4.5"	2.7"	6.25"
3.5"	4"	5.25"	3.25"	7.25"
4"	4.5"	6"	3.75"	8.25"
5"	5.563"	7.5"	4.718"	10.2815"
6"	6.625"	9"	5.687"	12.3125"
8"	8.625"	12"	7.687"	16.3125"
10"	10.75"	15"	9.625"	20.375"
12"	12.75"	18"	11.62"	24.375"
14"	14"	21"	14"	28"
16"	16"	24"	16"	32"
18"	18"	27"	18"	36"
20"	20"	30"	20"	40"
22"	22"	33"	22"	44"
24"	24"	36"	24"	48"
26"	26"	39"	26"	52"
28"	28"	42"	28"	56"
30"	30"	45"	30"	60"
32"	32"	48"	32"	64"
34"	34"	51"	34"	68"
36"	36"	54"	36"	72"
42"	42"	63"	42"	84"

The Shortest Combination Offsets

This chart is a guide for finding the combination of elbows needed to make the smallest combination offset possible using a 90° elbow and an odd angle elbow. If the nominal pipe size is divided into the offset needed and compared to the factors shown, the results will be the angle of the elbow to use in combination with a 90° ell to achieve the needed offset.

θ ell+ 90°ell	Offset Factor	θ ell+ 90°ell	Offset Factor	θ ell+ 90°ell	Offset Factor
1°	0.0264	31°	0.9868	61°	2.0847
2°	0.0533	32°	1.0228	62°	2.1202
3°	0.0806	33°	1.0590	63°	2.1555
4°	0.1083	34°	1.0952	64°	2.1906
5°	0.1364	35°	1.1316	65°	2.2255
6°	0.1650	36°	1.1682	66°	2.2602
7°	0.1940	37°	1.2048	67°	2.2947
8°	0.2234	38°	1.2415	68°	2.3289
9°	0.2531	39°	1.2783	69°	2.3628
10°	0.2833	40°	1.3151	70°	2.3965
11°	0.3138	41°	1.3520	71°	2.4299
12°	0.3446	42°	1.3890	72°	2.4631
13°	0.3759	43°	1.4260	73°	2.4959
14°	0.4074	44°	1.4630	74°	2.5284
15°	0.4393	45°	1.5000	75°	2.5607
16°	0.4716	46°	1.5370	76°	2.5926
17°	0.5041	47°	1.5740	77°	2.6241
18°	0.5369	48°	1.6110	78°	2.6554
19°	0.5701	49°	1.6480	79°	2.6862
20°	0.6035	50°	1.6849	80°	2.7167
21°	0.6372	51°	1.7217	81°	2.7469
22°	0.6711	52°	1.7585	82°	2.7766
23°	0.7053	53°	1.7952	83°	2.8060
24°	0.7398	54°	1.8318	84°	2.8350
25°	0.7745	55°	1.8684	85°	2.8636
26°	0.8094	56°	1.9048	86°	2.8917
27°	0.8445	57°	1.9410	87°	2.9194
28°	0.8798	58°	1.9772	88°	2.9467
29°	0.9153	59°	2.0132	89°	2.9736
30°	0.9510	60°	2.0490	90°	3.0000
θ ell+ 90°ell	Offset Factor	θ ell+ 90°ell	Offset Factor	θ ell+ 90°ell	Offset Factor

This chart cannot be used with the ½" or ¾" elbows.

The Shortest Simple Offsets

Ang	Offset	Ang	Offset	Ang	Offset	Ang	Offset
0.5°	0.0001	23°	0.2384	45.5°	0.8973	68°	1.8761
1°	0.0005	23.5°	0.2488	46°	0.9160	68.5°	1.9005
1.5°	0.0010	24°	0.2594	46.5°	0.9349	69°	1.9249
2°	0.0018	24.5°	0.2701	47°	0.9540	69.5°	1.9493
2.5°	0.0029	25°	0.2811	47.5°	0.9732	70°	1.9739
3°	0.0041	25.5°	0.2922	48°	0.9926	70.5°	1.9985
3.5°	0.0056	26°	0.3036	48.5°	1.0121	71°	2.0233
4°	0.0073	26.5°	0.3152	49°	1.0318	71.5°	2.0480
4.5°	0.0092	27°	0.3270	49.5°	1.0516	72°	2.0729
5°	0.0114	27.5°	0.3390	50°	1.0716	72.5°	2.0978
5.5°	0.0138	28°	0.3512	50.5°	1.0917	73°	2.1228
6°	0.0164	28.5°	0.3635	51°	1.1120	73.5°	2.1479
6.5°	0.0193	29°	0.3761	51.5°	1.1324	74°	2.1730
7°	0.0224	29.5°	0.3889	52°	1.1530	74.5°	2.1982
7.5°	0.0257	30°	0.4020	52.5°	1.1737	75°	2.2235
8°	0.0292	30.5°	0.4151	53°	1.1945	75.5°	2.2488
8.5°	0.0330	31°	0.4285	53.5°	1.2155	76°	2.2742
9°	0.0369	31.5°	0.4421	54°	1.2366	76.5°	2.2996
9.5°	0.0411	32°	0.4558	54.5°	1.2579	77°	2.3251
10°	0.0456	32.5°	0.4698	55°	1.2792	77.5°	2.3506
10.5°	0.0502	33°	0.4840	55.5°	1.3008	78°	2.3762
11°	0.0551	33.5°	0.4983	56°	1.3224	78.5°	2.4018
11.5°	0.0602	34°	0.5129	56.5°	1.3442	79°	2.4275
12°	0.0656	34.5°	0.5276	57°	1.3661	79.5°	2.4532
12.5°	0.0711	35°	0.5425	57.5°	1.3881	80°	2.4790
13°	0.0769	35.5°	0.5576	58°	1.4102	80.5°	2.5048
13.5°	0.0829	36°	0.5729	58.5°	1.4325	81°	2.5306
14°	0.0891	36.5°	0.5884	59°	1.4549	81.5°	2.5565
14.5°	0.0956	37°	0.6041	59.5°	1.4774	82°	2.5824
15°	0.1022	37.5°	0.6199	60°	1.5	82.5°	2.6084
15.5°	0.1091	38°	0.6360	60.5°	1.5227	83°	2.6343
16°	0.1162	38.5°	0.6522	61°	1.5455	83.5°	2.6603
16.5°	0.1235	39°	0.6685	61.5°	1.5685	84°	2.6863
17°	0.1311	39.5°	0.6851	62°	1.5916	84.5°	2.7124
17.5°	0.1388	40°	0.7019	62.5°	1.6147	85°	2.7385
18°	0.1468	40.5°	0.7188	63°	1.6380	85.5°	2.7646
18.5°	0.1550	41°	0.7359	63.5°	1.6614	86°	2.7907
19°	0.1634	41.5°	0.7531	64°	1.6849	86.5°	2.8168
19.5°	0.1721	42°	0.7706	64.5°	1.7084	87°	2.8429
20°	0.1809	42.5°	0.7881	65°	1.7321	87.5°	2.8691
20.5°	0.1900	43°	0.8059	65.5°	1.7559	88°	2.8952
21°	0.1993	43.5°	0.8239	66°	1.7798	88.5°	2.9214
21.5°	0.2087	44°	0.8420	66.5°	1.8037	89°	2.9476
22°	0.2184	44.5°	0.8602	67°	1.8278	89.5°	2.9737
22.5°	0.2284	45°	0.8787	67.5°	1.8519	90°	2.9999

On every job there is always a run in which a pipe needs to be offset just a short distance to line up with a pump or the opening in a vessel. To determine the angle of the elbows needed for the offset you can divide the distance of the offset by the nominal pipe size and compare the answer to these offset factors. For example, you have a 12" line of pipe that needs to be offset $4\frac{3}{8}$".

$$\frac{4.375}{12} = .3646$$

The closest offset factor is the factor for 28.5°. That means that two 12" 28.5° elbows welded together will offset the line about $4\frac{3}{8}$".

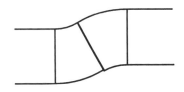

Functions Table

Deg ↓	Radian↓	Sin θ↓	Cos θ↓	Tan θ↓	Cot θ↓	Sec θ ↓	Csc θ↓		
0°	0.0000	0.0000	1.0000	0.0000	-	1.0000	-	1.5708	90°
0.5°	0.0087	0.0087	1.0000	0.0087	114.589	1.0000	114.593	1.5621	89.5°
1°	0.0175	0.0175	0.9998	0.0175	57.2900	1.0002	57.2987	1.5533	89°
1.5°	0.0262	0.0262	0.9997	0.0262	38.1885	1.0003	38.2016	1.5446	88.5°
2°	0.0349	0.0349	0.9994	0.0349	28.6363	1.0006	28.6537	1.5359	88°
2.5°	0.0436	0.0436	0.9990	0.0437	22.9038	1.0010	22.9256	1.5272	87.5°
3°	0.0524	0.0523	0.9986	0.0524	19.0811	1.0014	19.1073	1.5184	87°
3.5°	0.0611	0.0610	0.9981	0.0612	16.3499	1.0019	16.3804	1.5097	86.5°
4°	0.0698	0.0698	0.9976	0.0699	14.3007	1.0024	14.3356	1.5010	86°
4.5°	0.0785	0.0785	0.9969	0.0787	12.7062	1.0031	12.7455	1.4923	85.5°
5°	0.0873	0.0872	0.9962	0.0875	11.4301	1.0038	11.4737	1.4835	85°
5.5°	0.0960	0.0958	0.9954	0.0963	10.3854	1.0046	10.4334	1.4748	84.5°
6°	0.1047	0.1045	0.9945	0.1051	9.5144	1.0055	9.5668	1.4661	84°
6.5°	0.1134	0.1132	0.9936	0.1139	8.7769	1.0065	8.8337	1.4573	83.5°
7°	0.1222	0.1219	0.9925	0.1228	8.1443	1.0075	8.2055	1.4486	83°
7.5°	0.1309	0.1305	0.9914	0.1317	7.5958	1.0086	7.6613	1.4399	82.5°
8°	0.1396	0.1392	0.9903	0.1405	7.1154	1.0098	7.1853	1.4312	82°
8.5°	0.1484	0.1478	0.9890	0.1495	6.6912	1.0111	6.7655	1.4224	81.5°
9°	0.1571	0.1564	0.9877	0.1584	6.3138	1.0125	6.3925	1.4137	81°
9.5°	0.1658	0.1650	0.9863	0.1673	5.9758	1.0139	6.0589	1.4050	80.5°
10°	0.1745	0.1736	0.9848	0.1763	5.6713	1.0154	5.7588	1.3963	80°
10.5°	0.1833	0.1822	0.9833	0.1853	5.3955	1.0170	5.4874	1.3875	79.5°
11°	0.1920	0.1908	0.9816	0.1944	5.1446	1.0187	5.2408	1.3788	79°
11.5°	0.2007	0.1994	0.9799	0.2035	4.9152	1.0205	5.0159	1.3701	78.5°
12°	0.2094	0.2079	0.9781	0.2126	4.7046	1.0223	4.8097	1.3614	78°
12.5°	0.2182	0.2164	0.9763	0.2217	4.5107	1.0243	4.6202	1.3526	77.5°
13°	0.2269	0.2250	0.9744	0.2309	4.3315	1.0263	4.4454	1.3439	77°
13.5°	0.2356	0.2334	0.9724	0.2401	4.1653	1.0284	4.2837	1.3352	76.5°
14°	0.2443	0.2419	0.9703	0.2493	4.0108	1.0306	4.1336	1.3265	76°
14.5°	0.2531	0.2504	0.9681	0.2586	3.8667	1.0329	3.9939	1.3177	75.5°
15°	0.2618	0.2588	0.9659	0.2679	3.7321	1.0353	3.8637	1.3090	75°
15.5°	0.2705	0.2672	0.9636	0.2773	3.6059	1.0377	3.7420	1.3003	74.5°
16°	0.2793	0.2756	0.9613	0.2867	3.4874	1.0403	3.6280	1.2915	74°
16.5°	0.2880	0.2840	0.9588	0.2962	3.3759	1.0429	3.5209	1.2828	73.5°
17°	0.2967	0.2924	0.9563	0.3057	3.2709	1.0457	3.4203	1.2741	73°
17.5°	0.3054	0.3007	0.9537	0.3153	3.1716	1.0485	3.3255	1.2654	72.5°
18°	0.3142	0.3090	0.9511	0.3249	3.0777	1.0515	3.2361	1.2566	72°
18.5°	0.3229	0.3173	0.9483	0.3346	2.9887	1.0545	3.1515	1.2479	71.5°
19°	0.3316	0.3256	0.9455	0.3443	2.9042	1.0576	3.0716	1.2392	71°
19.5°	0.3403	0.3338	0.9426	0.3541	2.8239	1.0608	2.9957	1.2305	70.5°
20°	0.3491	0.3420	0.9397	0.3640	2.7475	1.0642	2.9238	1.2217	70°
20.5°	0.3578	0.3502	0.9367	0.3739	2.6746	1.0676	2.8555	1.2130	69.5°
21°	0.3665	0.3584	0.9336	0.3839	2.6051	1.0711	2.7904	1.2043	69°
21.5°	0.3752	0.3665	0.9304	0.3939	2.5386	1.0748	2.7285	1.1956	68.5°
22°	0.3840	0.3746	0.9272	0.4040	2.4751	1.0785	2.6695	1.1868	68°
		Cos θ ↑	Sin θ ↑	Cot θ ↑	Tan θ ↑	Csc θ ↑	Sec θ ↑	Radian↑	↑Deg

Functions Table

Deg↓	Radian↓	Sin θ ↓	Cos θ ↓	Tan θ ↓	Cot θ ↓	Sec θ ↓	Csc θ ↓		
22.5°	0.3927	0.3827	0.9239	0.4142	2.4142	1.0824	2.6131	1.1781	67.5°
23°	0.4014	0.3907	0.9205	0.4245	2.3559	1.0864	2.5593	1.1694	67°
23.5°	0.4102	0.3987	0.9171	0.4348	2.2998	1.0904	2.5078	1.1606	66.5°
24°	0.4189	0.4067	0.9135	0.4452	2.2460	1.0946	2.4586	1.1519	66°
24.5°	0.4276	0.4147	0.9100	0.4557	2.1943	1.0989	2.4114	1.1432	65.5°
25°	0.4363	0.4226	0.9063	0.4663	2.1445	1.1034	2.3662	1.1345	65°
25.5°	0.4451	0.4305	0.9026	0.4770	2.0965	1.1079	2.3228	1.1257	64.5°
26°	0.4538	0.4384	0.8988	0.4877	2.0503	1.1126	2.2812	1.1170	64°
26.5°	0.4625	0.4462	0.8949	0.4986	2.0057	1.1174	2.2412	1.1083	63.5°
27°	0.4712	0.4540	0.8910	0.5095	1.9626	1.1223	2.2027	1.0996	63°
27.5°	0.4800	0.4617	0.8870	0.5206	1.9210	1.1274	2.1657	1.0908	62.5°
28°	0.4887	0.4695	0.8829	0.5317	1.8807	1.1326	2.1301	1.0821	62°
28.5°	0.4974	0.4772	0.8788	0.5430	1.8418	1.1379	2.0957	1.0734	61.5°
29°	0.5061	0.4848	0.8746	0.5543	1.8040	1.1434	2.0627	1.0647	61°
29.5°	0.5149	0.4924	0.8704	0.5658	1.7675	1.1490	2.0308	1.0559	60.5°
30°	0.5236	0.5000	0.8660	0.5774	1.7321	1.1547	2.0000	1.0472	60°
30.5°	0.5323	0.5075	0.8616	0.5890	1.6977	1.1606	1.9703	1.0385	59.5°
31°	0.5411	0.5150	0.8572	0.6009	1.6643	1.1666	1.9416	1.0297	59°
31.5°	0.5498	0.5225	0.8526	0.6128	1.6319	1.1728	1.9139	1.0210	58.5°
32°	0.5585	0.5299	0.8480	0.6249	1.6003	1.1792	1.8871	1.0123	58°
32.5°	0.5672	0.5373	0.8434	0.6371	1.5697	1.1857	1.8612	1.0036	57.5°
33°	0.5760	0.5446	0.8387	0.6494	1.5399	1.1924	1.8361	0.9948	57°
33.5°	0.5847	0.5519	0.8339	0.6619	1.5108	1.1992	1.8118	0.9861	56.5°
34°	0.5934	0.5592	0.8290	0.6745	1.4826	1.2062	1.7883	0.9774	56°
34.5°	0.6021	0.5664	0.8241	0.6873	1.4550	1.2134	1.7655	0.9687	55.5°
35°	0.6109	0.5736	0.8192	0.7002	1.4281	1.2208	1.7434	0.9599	55°
35.5°	0.6196	0.5807	0.8141	0.7133	1.4019	1.2283	1.7221	0.9512	54.5°
36°	0.6283	0.5878	0.8090	0.7265	1.3764	1.2361	1.7013	0.9425	54°
36.5°	0.6370	0.5948	0.8039	0.7400	1.3514	1.2440	1.6812	0.9338	53.5°
37°	0.6458	0.6018	0.7986	0.7536	1.3270	1.2521	1.6616	0.9250	53°
37.5°	0.6545	0.6088	0.7934	0.7673	1.3032	1.2605	1.6427	0.9163	52.5°
38°	0.6632	0.6157	0.7880	0.7813	1.2799	1.2690	1.6243	0.9076	52°
38.5°	0.6720	0.6225	0.7826	0.7954	1.2572	1.2778	1.6064	0.8988	51.5°
39°	0.6807	0.6293	0.7771	0.8098	1.2349	1.2868	1.5890	0.8901	51°
39.5°	0.6894	0.6361	0.7716	0.8243	1.2131	1.2960	1.5721	0.8814	50.5°
40°	0.6981	0.6428	0.7660	0.8391	1.1918	1.3054	1.5557	0.8727	50°
40.5°	0.7069	0.6494	0.7604	0.8541	1.1708	1.3151	1.5398	0.8639	49.5°
41°	0.7156	0.6561	0.7547	0.8693	1.1504	1.3250	1.5243	0.8552	49°
41.5°	0.7243	0.6626	0.7490	0.8847	1.1303	1.3352	1.5092	0.8465	48.5°
42°	0.7330	0.6691	0.7431	0.9004	1.1106	1.3456	1.4945	0.8378	48°
42.5°	0.7418	0.6756	0.7373	0.9163	1.0913	1.3563	1.4802	0.8290	47.5°
43°	0.7505	0.6820	0.7314	0.9325	1.0724	1.3673	1.4663	0.8203	47°
43.5°	0.7592	0.6884	0.7254	0.9490	1.0538	1.3786	1.4527	0.8116	46.5°
44°	0.7679	0.6947	0.7193	0.9657	1.0355	1.3902	1.4396	0.8029	46°
44.5°	0.7767	0.7009	0.7133	0.9827	1.0176	1.4020	1.4267	0.7941	45.5°
45°	0.7854	0.7071	0.7071	1.0000	1.0000	1.4142	1.4142	0.7854	45°
		Cos θ ↑	Sin θ ↑	Cot θ ↑	Tan θ ↑	Csc θ ↑	Sec θ ↑	Radian↑	↑Deg

Index

Notes

Notes

Notes

Notes

Pipe Fitter's Math Guide
Quick Reference

$$a^2 + b^2 = c^2$$

$$\sqrt{a^2 + b^2} = c$$

$$\sqrt{c^2 - a^2} = b \text{ or } \sqrt{c^2 - b^2} = a$$

Sine θ	$= \dfrac{\text{Opposite Side}}{\text{Hypotenuse}}$	Cosecant θ	$= \dfrac{\text{Hypotenuse}}{\text{Opposite Side}}$	
Cosine θ	$= \dfrac{\text{Adjacent Side}}{\text{Hypotenuse}}$	Secant θ	$= \dfrac{\text{Hypotenuse}}{\text{Adjacent Side}}$	
Tangent θ	$= \dfrac{\text{Opposite Side}}{\text{Adjacent Side}}$	Cotangent θ	$= \dfrac{\text{Opposite Side}}{\text{Adjacent Side}}$	

Sine θ	$= \dfrac{1}{\text{Cosecant } \theta}$	Cosecant θ	$= \dfrac{1}{\text{Sine } \theta}$	
Cosine θ	$= \dfrac{1}{\text{Secant } \theta}$	Secant θ	$= \dfrac{1}{\text{Cosine } \theta}$	
Tangent θ	$= \dfrac{1}{\text{Cotangent } \theta}$	Cotangent θ	$= \dfrac{1}{\text{Tangent } \theta}$	

Hypotenuse	=	Cosecant θ	x	Opposite side
Hypotenuse	=	Secant θ	x	Adjacent side
Opposite side	=	Tangent θ	x	Adjacent side
Opposite side	=	Sine θ	x	Hypotenuse
Adjacent side	=	Cosine θ	x	Hypotenuse
Adjacent side	=	Cotangent θ	x	Opposite side

Pi $(\pi) \approx 3.1416$

Circumference $= \pi d$

Circumference $= 2\pi r$

Area of circle $= \pi r^2$

Volume of cylinder $= \pi r^2$ x length

Radians $= \theta \text{ x } \dfrac{\pi}{180}$

Arc length = radians x radius

Chord length $= \sin \dfrac{\theta}{2}$ x radius x 2 or Chord length $= \sin \dfrac{\theta}{2}$ x diameter

Take out $= \tan \dfrac{\theta}{2}$ x radius of elbow

$\sqrt{2} \approx 1.4142$